Suspicious Tech
By: Mina G.
USA 2018

Copyrights

Table of Contents

To the love of my life, Mariem ~

Without her encouragement, this book would have

been completed five years later

Thank you

<u>Angie F.</u>   You are my honest friend who I deeply appreciate and cherish. Having you as a friend means everything. I love you to the moon and back.

<u>Igor Benić</u>   Thank you for the great plugin "wpsimplegiveaways" I enjoyed using it! To check Igor's website visit www.ibenic.com

I was born in Giza/Egypt. The city where Egypt's most famous monuments "The Pyramids" exists. At the age of 26, I immigrated to the USA after graduation from college with a bachelor's degree in Civil Engineering. As a new immigrant, it was very difficult to find a job remotely related to my degree. Therefore, I had to work in almost every trade to afford to live in my new home, where the dreams come true; the USA. Starting from a customer service agent over the phone ending with an Amazon Flex driver. Among other professions, I was a wedding photographer, a security system installation technician and eventually a civil engineer. During all of that, I started my photography channel on YouTube in 2015. Basically, I had to use everything I am experienced in to stay alive. Back in Egypt, I had my small web hosting/graphic design company which gave me enough experience to design websites, edit videos and also practice photography.

With such experience in photography and digital content and surveillance devices, I was often asked what a spy camera looks like nowadays and is it something similar to what thriller and espionage movies usually show?! My answer was always, "They can come in every shape you imagine." Sometimes I would spend hours with a friend explaining how to spot hidden cameras easily. And after a while of doing so, I decided to put everything I know in one book and share it with everyone. And that's how Suspicious Tech came to life.

The book is intended to be the conclusion of 12 years of experience and master reference for everything related to identifying spy cameras nowadays.

Introduction & How the book works

Thank you for buying the **Suspicious Tech** book. I hope you find it useful in spotting hidden cameras anywhere. My name is Mina, and I am the author of this book. I have been practicing photography, videography, web design and installing security camera systems for over 12 years. During my many years of experience, I have encountered almost every surveillance camera you can imagine. I was also a technical support supervisor for one of the largest camera companies in the world.

This book is divided into logical chapters. However, if you feel that you have enough knowledge about one of the subjects, feel free to skip that section and to read the specific chapter you are interested in. The section everyone usually skips is (Chapter One) the basics. So, if you feel confident in your videography knowledge or in other words, if you have ever used a video camera, feel free to skip Chapter One.

At the end of this book, you will find a unique code to unlock a bonus section on the book's official website. That bonus section contains exclusive video tutorials, tips and tricks all related to enhancing your security and privacy. Please find the code on the last page, and visit the bonus section on www.suspicious.tech

Also, under the bonus section, you can request a PDF copy of the book signed by me and sent to your email. All you need to do is to visit the bonus section URL and to use the form to send me your name, and I will send you a signed PDF copy of the book with colors (Kindle devices show any book in black and white only. However, Kindle app for Android and iOS can show colors).

The hyperlinks in this book will open normally if you are reading on Kindle App for Android or Kindle app for iOS. However, on Kindle devices, they will open very slowly like any other link because Kindle devices are not meant to open internet pages and they have limited internet browsing capabilities. If you are reading the book on Kindle device, I would highly recommend that you visit the book's official website to access the bonus section and to read articles and videos on the website directly. This book is meant to be an all-in-one reference to every hidden/spy camera available on the market today. Therefore, we won't cover old style DVR cameras because they simply no longer exist.

Chapter One
"The Basics"

1.1 Main Components

In order to spot hidden cameras around you, you will need to understand the basic parts of any camera and the function every part plays. In this chapter, you will learn the basics which will help you understand the following chapters.

You can easily recognize one major turn in the course of camera evolution by the time when digital videography ended the era of analog videography. Therefore, in this book, we will discuss digital video cameras only because technically no one is using analog video cameras.

A typical video camera consists of a lens in front of a light sensor. Both of the camera and the lens are attached to the same body (container), and in the same container, there is a processing unit (a device to translate all the information from the sensor and make the final video file). That device is called the camera processor. And eventually, the processor saves that video file to a storage unit such as a hard drive (usually referred to by HDD) or a flash drive (usually referred to by SD Card).

That was the most basic components of a functional video camera. Without any doubt, most cameras have more advanced components and more gimmicks into them for obvious reasons such as controlling the lighting, zooming in and out, panning, motion detection and even uploading the video to the internet... etc.

While not every camera has the same extras, they all share the same components mentioned above to record video footage.

What you need to know from this section: A typical camera consists of a lens, light sensor, processor and a storage unit.

1.2 The Lens

The lens used in surveillance cameras is usually a wide-angle lens, which means the lens can cover a larger area for obvious reasons; using a fewer number of cameras as possible. You can see the difference between wide angle and narrow-angle (also called telephoto) lens on the pictures below:

As you can see from the two pictures above, a wide-angle lens can broaden your field of view; however, a telephoto lens can focus on one specific subject/object.

In addition to the angle of view of each lens, the material it's made of makes a very big factor on how expensive the lens is hence, how expensive the final product (the camera) is.

Without going into too many details; some lenses are coded to deliver a clearer image to the camera sensor and eventually clearer video footage.

Also, some lenses are lighter than others depending on the material they are made of.

What you need to know from this section: The lenses used in surveillance video cameras are usually wide-angle lenses to cover as many people or objects in the video as possible.

1.3 The Light Sensor

The light sensor is the most expensive part of any camera because, most of the footage quality, depends on how good the sensor is. Have you ever come across the term "High Definition Video" or "HD Video"? It means the video produced is 1280 pixels X 720 pixels. Knowing that one pixel is the standard unit of measurement for video footage and digital images.

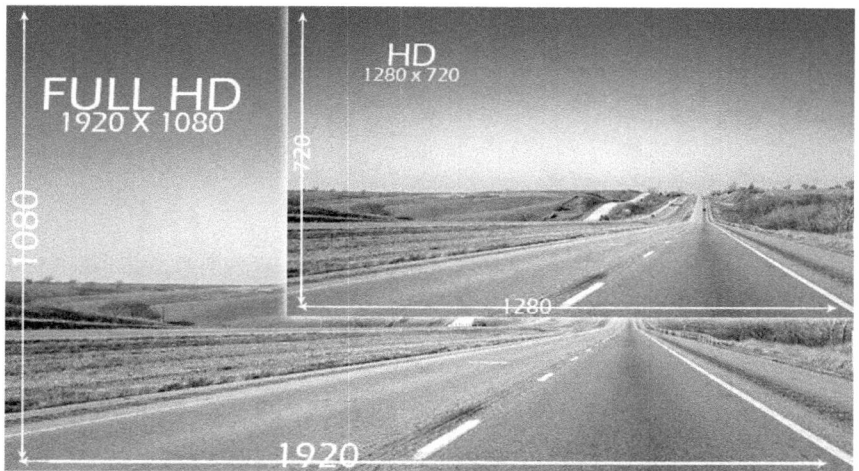

So, mainly every new video resolution gets a fancy name to give you the impression that it's bigger and better. For example, Ultra HD (also referred to by 4K) is 3840 x 2160 pixels. And 8K is 7680×4320 pixels.

That being said, most hidden cameras can produce HD resolution or below! Most of the cheap spy cameras sold on eBay are SD (Standard Definition) 640x480 pixels. That's even less in quality than HD.

The higher the resolution the camera sensor can produce, the more expensive the camera gets. Also, the larger the sensor, the better the footage quality will be in low light conditions because every pixel will absorb more light if one pixel is large enough.

1.4 The Memory Card

The memory card is the place where most cameras save the video footage. If you come from the era of videotapes, consider the memory card the alternative of the tape. And the memory card needs to be inside the camera during recording for the camera to save the footage on it. The memory card form factor is either SD (Secure Digital) or Micro SD (Micro Secure Digital) as shown below:

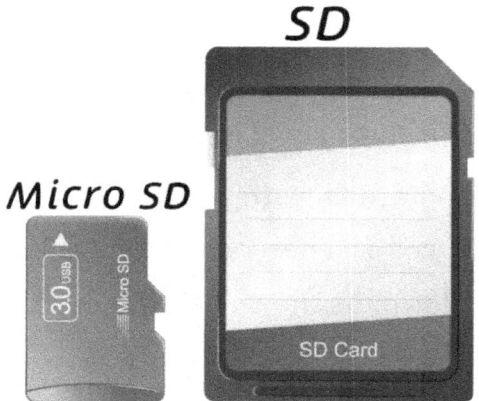

The only difference that matters is: one is larger in dimensions (SD), and the other is small. Because spy/hidden cameras need to be small, they usually use **Micro** SD Cards, instead of full-size SD.

The moment you spot a hidden camera that uses one of these memory cards, remove the memory card out of the camera and keep it with you for two reasons:

1. As an evidence that this place has a hidden camera.
2. To prevent the camera owner from using your video footage and from publishing your video online.

Keep in mind that having a spy/hidden camera installed in a public restroom or fitting room is illegal in the first place. Therefore, defending your invaded privacy by removing the storage unit (the SD card) (aka the memory card) shouldn't be considered a crime in any court.

To remove the Micro SD Card out of any camera, you need to push the card down, and the card will pop up out of the camera. Because the Micro SD Card is so tiny and the groove where it lives is also very tiny, you will need a small tip (like a pencil or so) to push the card down. As shown below:

A. You need to find out where the memory card lives. In the example below, you must remove the prongs first to expose the memory card slot:

 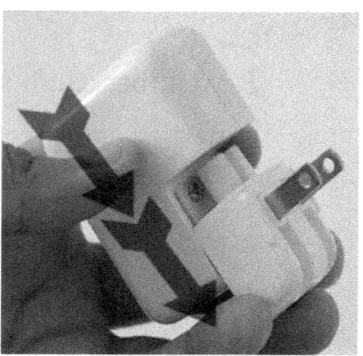

B. Then you need to push the memory card using a small tip such as the tip of a pen or a pencil:

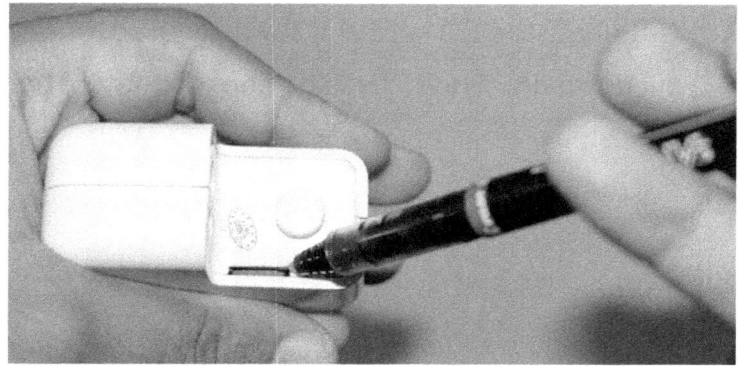

C. The memory card will pop up, and you will be able to remove it from the camera easily:

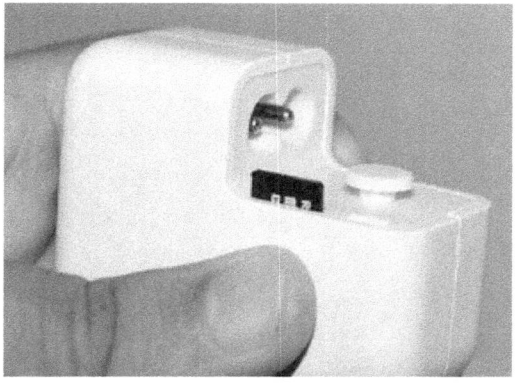

Remember that not every camera hides the memory card the same way. That's why it's important to read through Chapter Four "Indoor Object Cameras" to learn how each camera hides the memory card.

The general rule for hidden cameras is to hide the lens and the memory card. This way the victim wouldn't suspect a normal gadget to be a hidden camera. So, try to think outside the box. For example, a wall charger would hide the SD card slot towards the prongs because you normally won't see that part of the charger when connected to the AC outlet.

1.5 The Mysterious Cloud

On a movie called "Sex Tape" *Jason Segal* talking to *Cameron Diaz* describing a sex tape involving him and her, said, " It went up to the cloud!" She answered, "And you can't get it down from the cloud??" "Nobody understands the cloud; it's a mystery, " he said!

This scene was particularly funny to my generation because the word "Cloud" started to be used as a technical term slightly after our generation. A friend of mine once told me "The cloud is like a giant computer on gigantic wheels and wings to fly AND save data at the same time!

Now to the real world. The cloud consists of many computers connected to each other via a network, and they have redundant data saving tools which means if one of the hard drives (the devices to save data on) fails, another hard drive starts working simultaneously. The new hard drive will have the same data you lost on the first hard drive.

Example:
You have two identical hard drives: two Terabytes of space for each of them. Instead of using both and being able to store 4 Terabytes of data, you store only two Terabytes of data identically on each, in case one of them fails. What I just described is called "RAID."

Therefore, the cloud is a bunch of computers connected to the internet all over the world (not necessarily in the same place), and by combining many computers like that you gain more disk space and computing power, and everybody gets happier.

An example of the cloud is an online service offered by Google called "Google Drive" and its competitor from Microsoft is called "Microsoft One Drive" and another competitor would be BOX.COM, DROPBOX.COM and Amazon AWS.

They are all online services to give you some extra space to save your data (pictures, videos, text files, documents.. etc.) and not to worry if your local data (your hard drive connected to your PC or laptop) gets destroyed by a natural disaster, theft or even by accident.

It's very intriguing to pay for one of the mentioned above services to have a safe copy of all your important data saved and ready to be used the moment you are connected to the internet.

What we care about in this book is hidden cameras, and the bad news is, many hidden cameras available on the market are capable of saving video footage to the cloud the moment they record the video which means, even if you spot the hidden camera and destroy the camera, your private footage will live in the cloud until the camera owner download them and watch them and do god knows what with them.

Therefore, it's extremely important in the case of cloud cameras not to leave the scene until the authorities arrive because, if you leave and the camera owner manages to hide the camera you might not be able to track and stop your video footage from spreading over the internet plus you won't be able to link the camera owner to the video legally.

Many camera manufacturers offer cloud storage for the owners of their cameras with certain limitations such as how much space they can use, how many video clips they can keep or even how long each video will live in the cloud before it gets deleted.

What you need to know from this section: The question you need to ask yourself the moment you spot a hidden camera is "Where the footage is being saved?"

Because, if the camera is saving your private footage to the cloud, it won't help you to disconnect the camera from electricity or even if you destroy the actual camera. Your footage will be saved online, and the camera owner will have access to your footage anytime later by accessing the internet.

Chapter Two
"How They Hide?!"

For **someone** to videotape you, he/she needs to follow one of the two major approaches listed below;

1. Use a two-way mirror (one side reflective mirror and the other side see through glass). And, then they need to place a camera behind the see-through side. This approach is what we are going to discuss in Chapter Three "Two Way Mirror."

2. Place a regular object that looks completely harmless, such as a clock or a Bluetooth speaker, a bottle of water or even a cell phone charger which has a hidden camera in that harmless object. And this approach is going to be discussed in Chapter Four "Indoor Object Cameras."

Now let's be clear that these two approaches are the only available methods for "**someone**" to use a hidden camera, and that is not the case for an advanced intelligence agency such as the CIA. Let's agree that if the monitoring and surveillance techniques used by the CIA or any other intelligence agencies were made public, that would defeat the purpose, and that means someone at the CIA is not doing his job.

However, as an example of how this can be done without having an actual camera in place; there is a new technology to have heat-based video recordings using your WiFi signals coming from your internet router. Scary, isn't it? However, the good news is, that kind of advanced technology is not available to the public yet.

To see an experiment done by the University Of California Santa Barbara researchers in which the researchers were able to count how many people in a room using just wifi signals, please visit https://suspicious.tech/wificount

As mentioned before, this book will not discuss such extremely advanced methods of espionage as those advanced methods are not in the hands of the public. Therefore, you don't have to worry about it.

Chapter Three
"Two Way Mirror"

What is a two-way mirror and how it works?

A two-way mirror is just like a normal mirror from one side, while the other side is a see-through piece of glass.

In 2015 a lady found a two-way mirror in the ladies' restroom in a famous bar in Illinois, and the incident was all over the internet, and the news.

Here is the main problem with two-way mirrors; they are affordable and available on almost any major online store starting from $4 for the small-sized piece of mirror. The larger the mirror gets, the pricier it will be. On the next page, you can see a more detailed explanation of how that mirror works.

As you can see, one side looks exactly like a regular mirror:

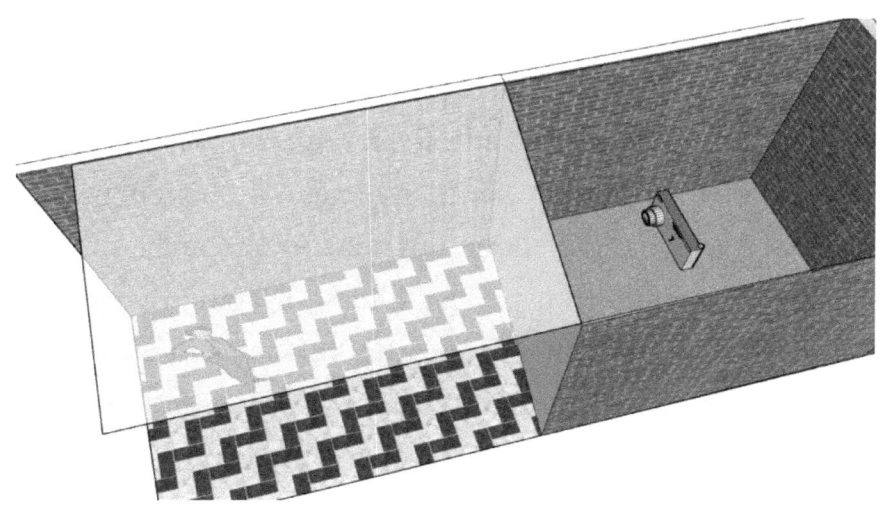

The other side is just a piece of see-through glass:

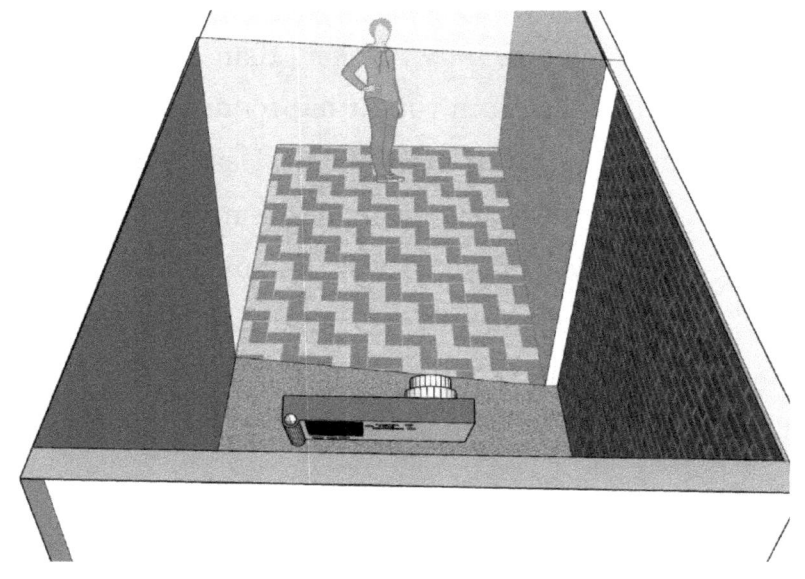

As you can see, it's a normal mirror on one side, and a semi-transparent glass from the other side. The theory of this mirror is: any surface that reflects more than 70% of the light is considered a mirror. So, they made a mirror with a reflection rate of 71% or above, but not 100% to make sure it's semi-reflective/semi-transparent. And by keeping the camera room dark and the fitting room well lit, the camera will see clearly from the two-way mirror, and the person in the fitting room won't see what's inside the camera room. He/she will simply see their reflection in the mirror.

However, for a two-way mirror to work, the camera room needs to be significantly darker than the fitting room. Otherwise, the person in the fitting room will be able to see through the two-way mirror and will see what's inside the camera room. I have made a small example to show you how having both sides of a two-way mirror well-lit will expose what's behind the mirror (the camera room) and, doing so defeats the purpose:

As you can see on the picture above, the part of the mirror that covers a dark room (the inside of the box is dark), works as a perfect mirror and reflects some of the books in my library on the opposite side of the mirror. However, the part of the mirror sticking out of the box and exposed to light from both sides is partially transparent. Hence, you can see the blue body of the toy through it. And, it's also partially reflective (you can see a reflection of one of the books on it also).

From all of that, we can conclude that the most effective way to find out whether the mirror in a fitting room is a two-way mirror or just a regular harmless mirror; is to turn the lights off in the fitting room and to point any flashlight such as your phone's flashlight towards the mirror. If you can see through the mirror, it's a two-way mirror.

That was the only accurate way to spot two-way mirrors. However, there are other methods, but they are not as accurate. That includes placing your finger on the mirror; if your fingernail touches its own reflection, then there is a 90% chance that you are looking at a two mirror.

After this method of detecting two-way mirrors became popular online on YouTube and other websites, many two-way mirror manufacturers started to use different materials that produces a similar reflection to the one produced by normal mirrors.

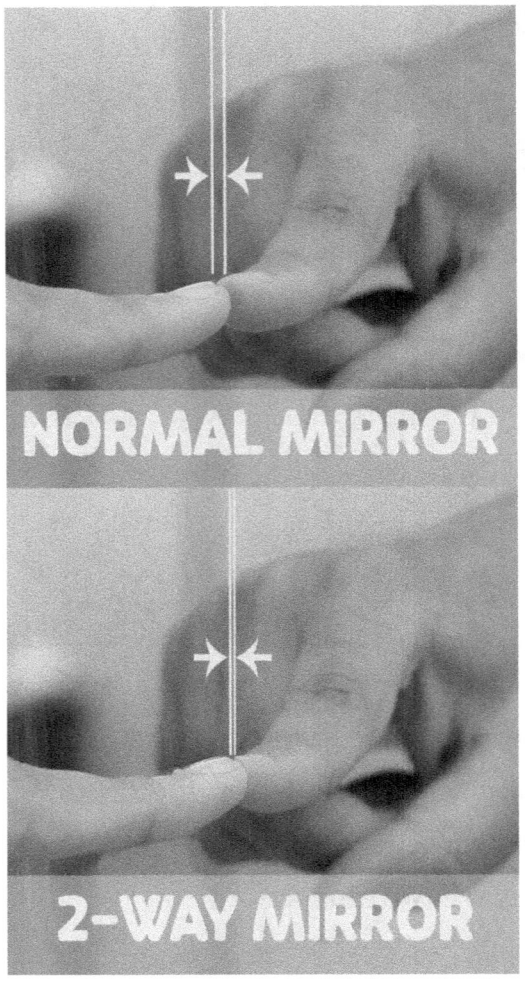

What you need to know from this section: To spot a two-way mirror you need to turn the lights off in the fitting room, and to put a flashlight against the mirror. If you can see the other side, it's a two-way mirror. Otherwise, it's a normal mirror.

Chapter Four
"Indoor Object Cameras"

Clock Camera#1

Although this clock looks harmless, it has a hidden camera on the left corner that can be seen only when you remove the front cover as you can see below:

This hidden camera has night vision and, connects to the internet via Wi-Fi to send alerts to the owner of the camera once motion is detected. The camera owner has the option to view action simultaneously or review it later as the camera saves videos to a Micro SD card that can be accessed by removing the back cover of the clock as shown below:

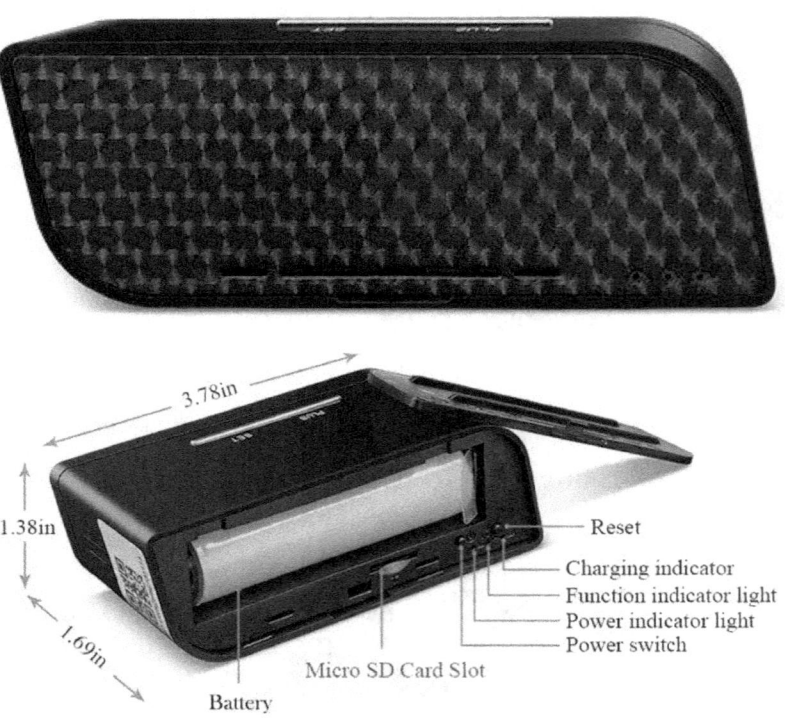

When you find this camera, remove the back cover and take the Micro SD Card with you as this specific model is not capable of uploading footage to the internet.

Therefore, it's safe to assume that your footage is not recorded elsewhere. Also, don't forget to call the authorities as using hidden cameras in any public restroom or fitting room (and in a private location without a clear notice saying so) is prohibited by the law almost everywhere.

And the same camera comes in other shapes loaded with the same user interface and features. Therefore, there is no need to list them all separately. The following is a list of the other shapes clock camera number one comes in:

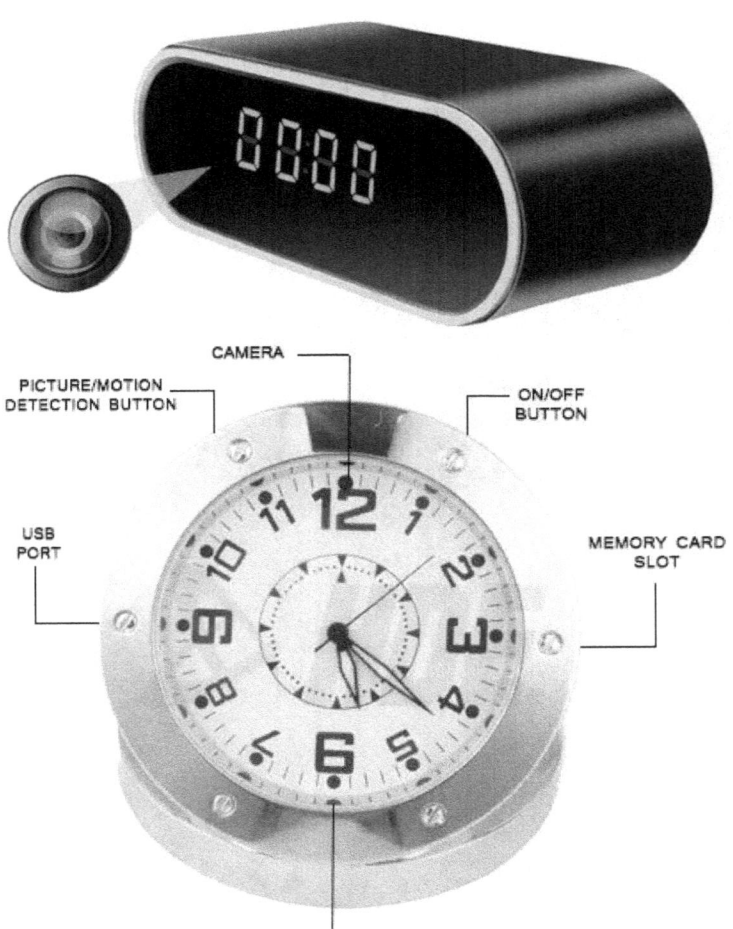

CAMERA

PICTURE/MOTION
DETECTION BUTTON

ON/OFF
BUTTON

USB
PORT

MEMORY CARD
SLOT

LIGHT INDICATOR

memory
card

Lens

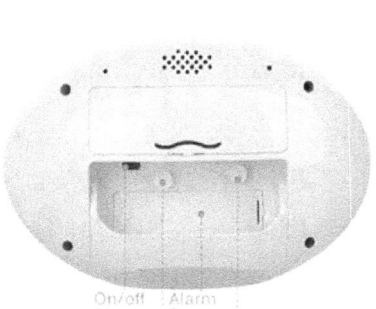
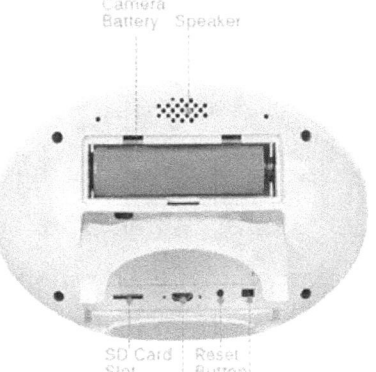

Camera
Battery   Speaker

On/off   Alarm
Alarm   Battery
Clock   Alarm   Time
         Clock   Adjust
         Setting   Button

SD Card   Reset
Slot   Button

Micro   On/off
USB   Camera
Port

Clock Camera#2

The camera on this clock lives right under the 6 o'clock (on the vertical dash under number 6), and this camera can store video footage on a Micro SD memory card. What's funny about this specific hidden camera is that the clock it comes behind is always branded "POWER." So, it's a good thing the makers of this specific model did not have enough creativity to change the fake brand! When you see this model, simply flip it around, and you will see the memory card (Micro SD Card) on the back of the camera without any particular way of hiding it.

However, it's under a small groove which means you will need a very small tip to get the card out of its place.

This camera is not capable of saving the footage to the cloud. Which means it saves to the local Micro SD Card only. However, it's capable of motion detection, and it sends an alert to the camera owner upon any motion detection. And the owner can liveview the motion if he/she decides but won't be able to re-watch the video until he gets his hands on the Micro SD Card inside the actual clock.

This one looks like a normal digital clock; however, the speaker hole is the lens, and it's well hidden and looks pretty normal to the average user. On the right side of the clock there is a small black door that you can move up and down to expose the memory card and remove it as shown below:

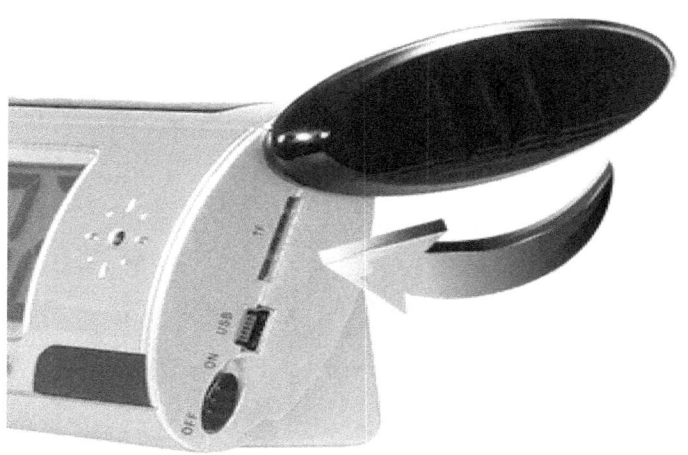

When you find this camera, simply move the side door up and remove the memory card out as this camera is capable of saving only to the local Micro SD Card and cannot save to the cloud. The memory card slot is the popup type. As you may already know by now; the popup type needs a very small tip to get the memory card to pop up.

Clock Camera#4

This is the cheap version of two cameras that have the same optics and features in two different cases. The other version is more expensive for no particular reason, and it's branded "NewWings." For this version, the lens is either on 6 or 10. The memory card lives in the back of the clock under a small sticker as shown below:

This camera is not capable of recording to the cloud. Therefore, you would be safe by taking the memory card out of the clock.

The other version is sold online under the brand "NewWings," and it looks slightly better with higher end finish, and the lens is way more difficult to find as you can see below:

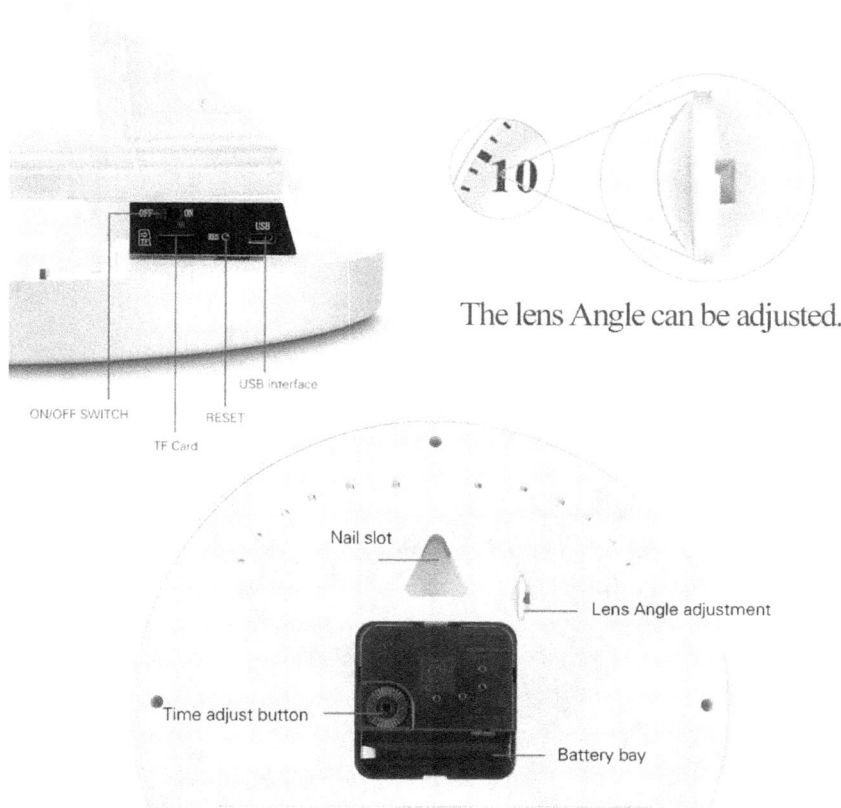

The lens Angle can be adjusted.

ON/OFF SWITCH

USB interface

RESET

TF Card

Nail slot

Lens Angle adjustment

Time adjust button

Battery bay

Clock Camera#5

8hrs
recording

64G
SD Card

 | | |  |

HD	Motion detection	Night vision	2 way communication	180° wide angle
1080P				

This hidden camera has a piece of two-way mirror to cover the lens. If you don't know what a two-way mirror is, please review Chapter Three.

That being said, it's very difficult to see the lens on this clock under normal lighting conditions as you can see on the picture above. And, to add insult to injury, it's also very difficult to find the memory card slot on this clock as you can see below the back of the clock is just a normal clock:

However, if you remove that black cover on the back of the clock, you will see the controls for the camera, and you will expose the memory card slot as shown

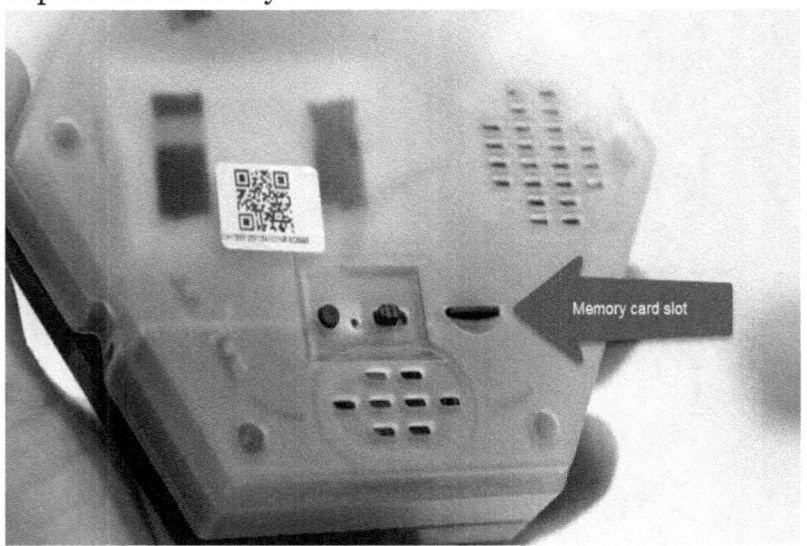

Memory card slot

Just a side note: this clock has a very nice looking side light that makes the clock a piece of art! So, it's very tempting for many people to use it as a good looking clock that can be used as a spy camera!

And more variations of the same idea:

Wall Charger Camera

Wall charger cameras are almost all the same. The side that goes into your AC outlet is where you can find all the ports, including the Micro SD Card slot. And, the other side is the camera lens as you can see in the following examples. Keep in mind that the more expensive model is available on this camera that can also record to the cloud.

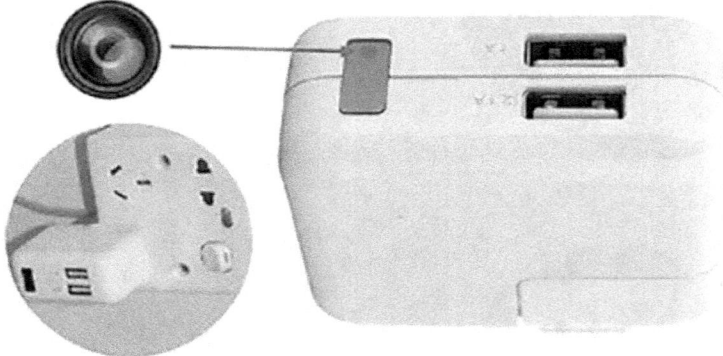

That means if you see this camera somewhere, you need to be 100% positive that it's the cheap version that that is not capable of recording to the cloud (the ones listed below), which you can easily neutralize by removing the memory card out of the camera.

A. To remove the memory card, you need to remove the prongs as shown below:

B. Then you need to push the memory card using a small tip such as the tip of a pen or a pencil:

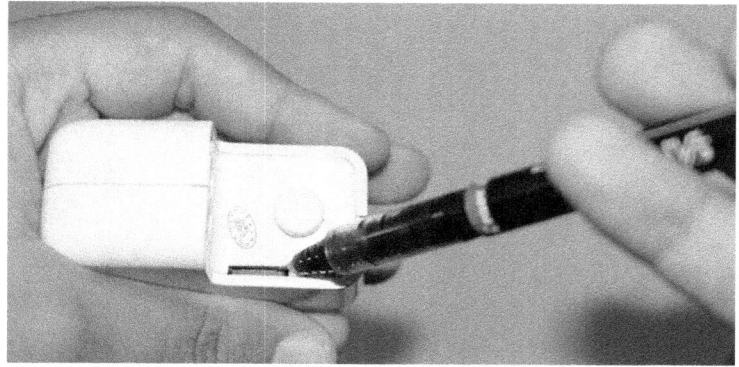

The memory card will pop up, and you will be able to remove it from the camera easily.

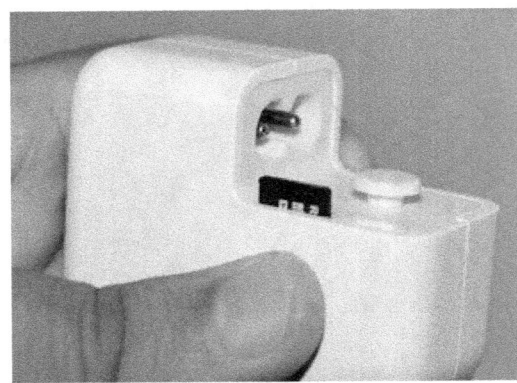

Another example of the same concept with a different location for the memory card. The memory card on this spy camera lives on top of the prongs.

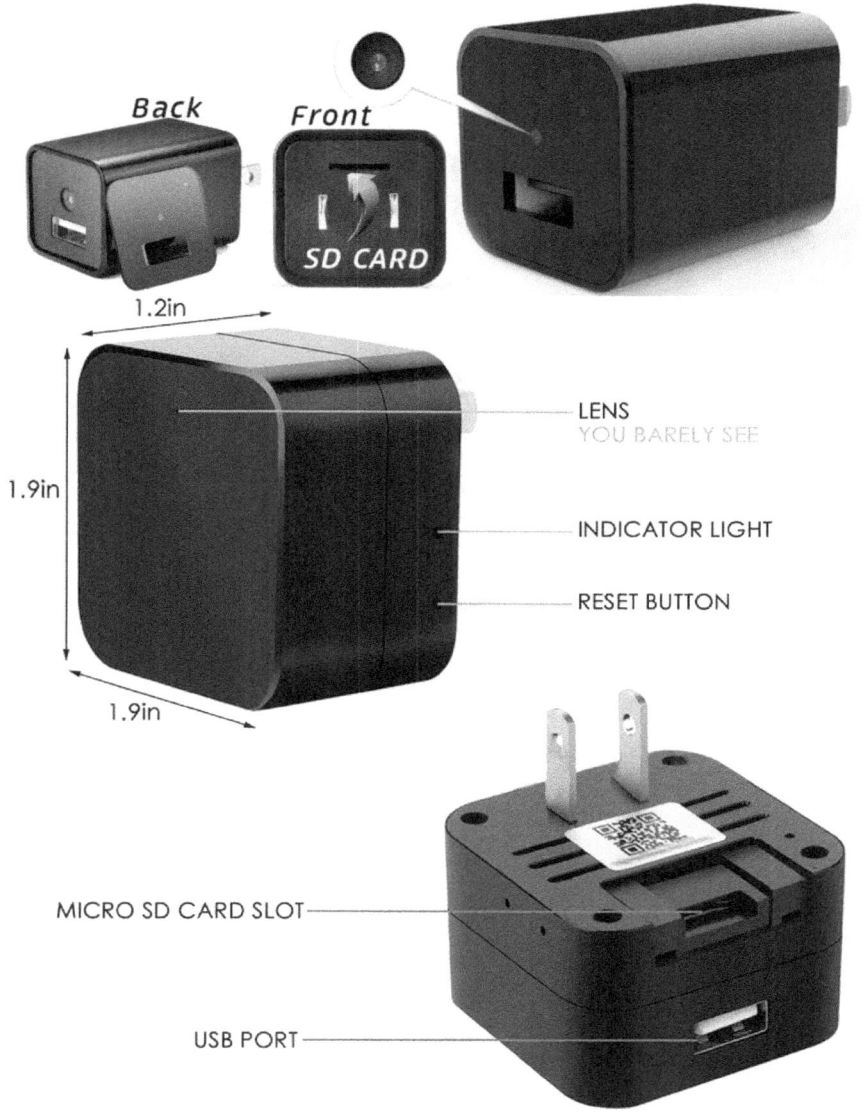

Clothes Hook "Hanger" Camera

This camera takes the shape of a simple clothes hanger as shown below

1 - VIDEO START/ STOP BUTTON   2 - CAMERA LENS

3 - RESET   4 - INDICATOR LIGHT

5 - CARD SLOT   6 - CHARGER PORT

7 - POWER ON / OFF

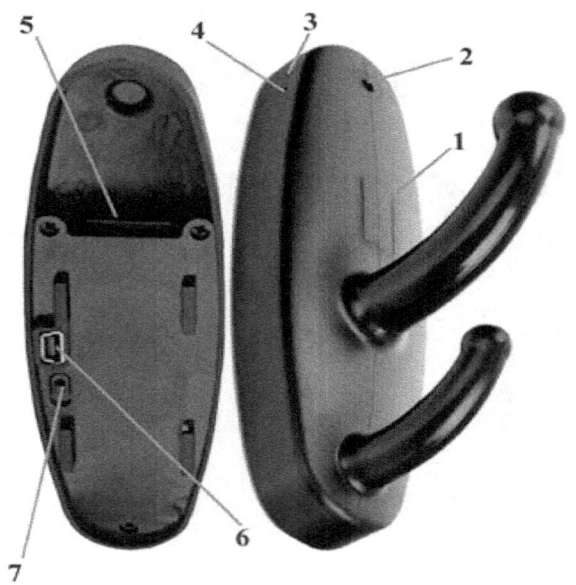

As you can see from the pictures above, the lens is located on the upper area of the clothes hook to make sure clothes do not cover it. This camera can record to a micro SD card inside the camera and is not capable of sending anything via WIFI. So, it's safe to assume that removing the SD card out of the camera is enough.

And another variation of the same idea:

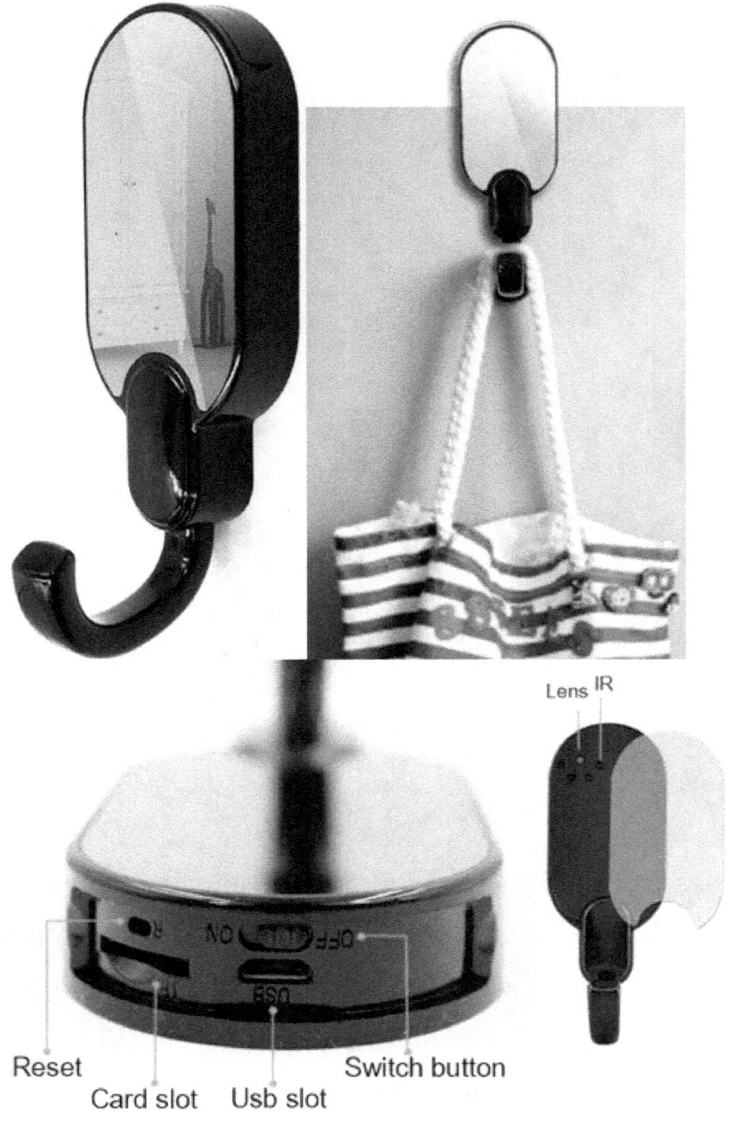

Lens IR

Reset

Card slot Usb slot

Switch button

Bluetooth Speaker Camera

It takes the shape of a normal shiny Bluetooth speaker, but don't be fooled; it's capable of much more than music streaming!

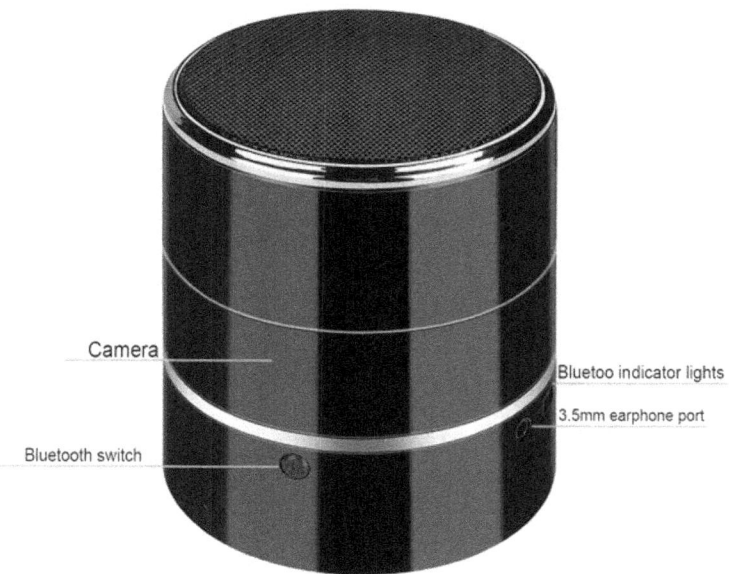

Camera

Bluetooth switch

Bluetoo indicator lights

3.5mm earphone port

The micro SD card is located at the bottom of this camera, so you will need to flip the Bluetooth speaker over to see the micro SD card slot:

Wi-Fi ON/OFF

Reset

Micro SD Slot

This is one of the higher end spy cameras. It's not one of the cheapo Chinese cameras you can get from eBay for $10. It starts from $80, and it can record 1080p (Full HD) which means things are getting serious here and you cannot assume that your face won't be recognized on the footage recorded by this spy camera.

This is one of the few cameras mentioned in this book that has a large enough battery to work without AC source for 5-6 hours, which again make it more dangerous than regular cheaply made spy cameras. The camera is capable of motion detection, and it's very sensitive according to the test we have made.

The camera sends an alert to the camera owner when motion is detected. The owner will receive the alert on his smartphone (Android and iPhone), and a screenshot of the first frame of motion will be saved to the camera owner's phone internal storage, and he/she will be able to live view the camera feed in real   time.

USB Charger Camera

This multiport USB charging station has a hidden camera in the front of the device

Wifi Camera

Motion detection

5 USB Ports

Charge multiple USB devices at the same time to meet your power demands with safe guarantee, even if the camera is monitoring.

5 USB

As you can see on the picture, the lens is in the upper area of the device right on top of the first USB charging port. The micro SD card chamber is on the back of the unit as you can see below:

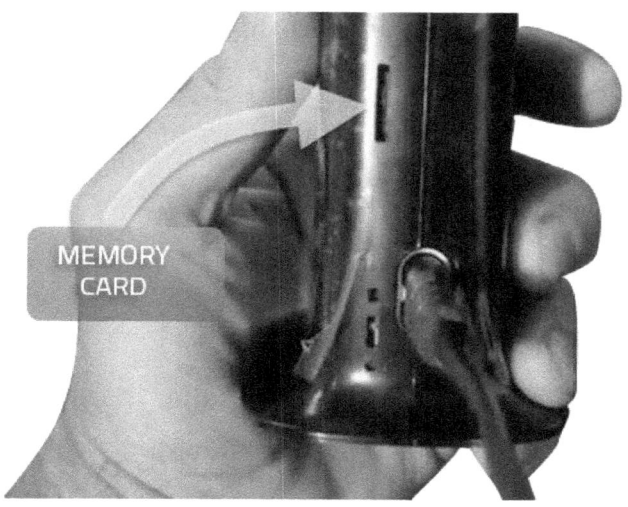

MEMORY
CARD

It's extremely difficult to remove the memory card out of this camera. Even the camera owners may find it difficult to remove the card out of this camera. So, if you can't remove the card, you may want to take the whole camera with you!

This camera is capable of streaming live video feed via Wi-Fi to the camera owner smartphone directly. And it sends notifications to the owner as upon motion detection.

Smoke Detector Camera

Smoke detector hidden are the most difficult hidden cameras to detect. The reason this camera is difficult to detect is the fact that this camera can hide in plain sight and you would never suspect it to be a spy camera!

Every facility should have a smoke detector! Right? But here is the catch: <u>Most state building codes recommend not installing a smoke alarm in or near a bathroom.</u>

So, if you see a smoke detector in a restroom, there is a 120% chance that you are looking at a hidden camera.

The other reason why this hidden camera is difficult to detect is being installed in the ceiling which you may not be able to reach without additional help such as using a ladder. So, you may suspect the smoke alarm to be a hidden camera, but you can't physically open the detector case to look inside the unit and see if there is a memory card slot or not! The simple rule of thumb is, if there are two smoke detectors in one room, one of them is a camera!

WIFI
Camera

HD Video

Motion
Detection

TypeI: Fixed lens – Looking straight down

This type of fake smoke detectors can look only downwards, and the angle covered by this hidden camera is the area exactly under the smoke detector as shown below:

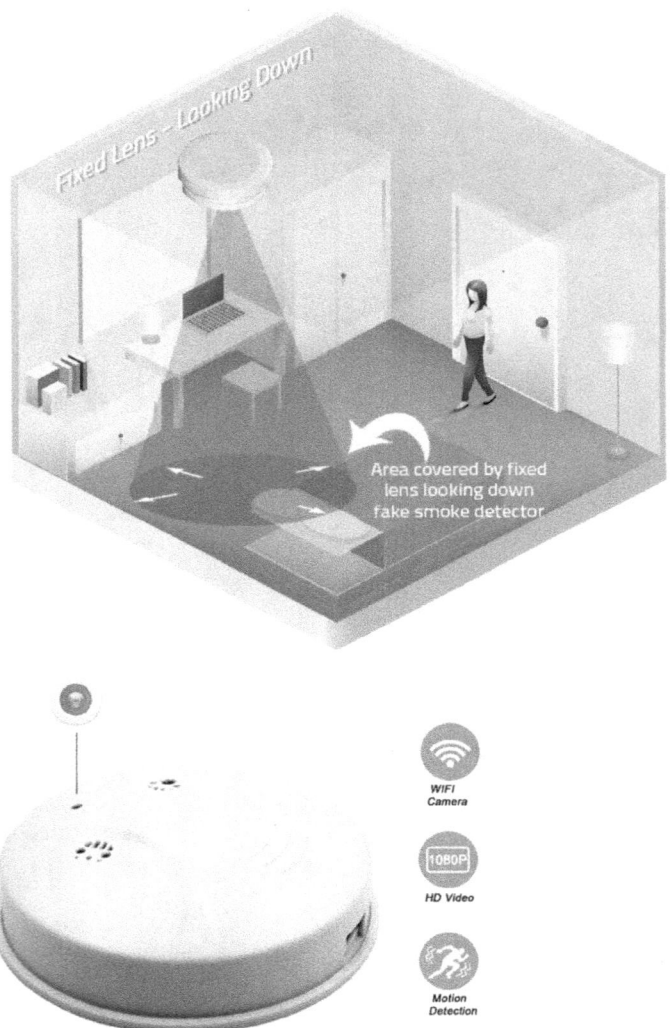

Area covered by fixed lens looking down fake smoke detector

WIFI Camera

HD Video

Motion Detection

Type II: Fixed lens – Tilted angle

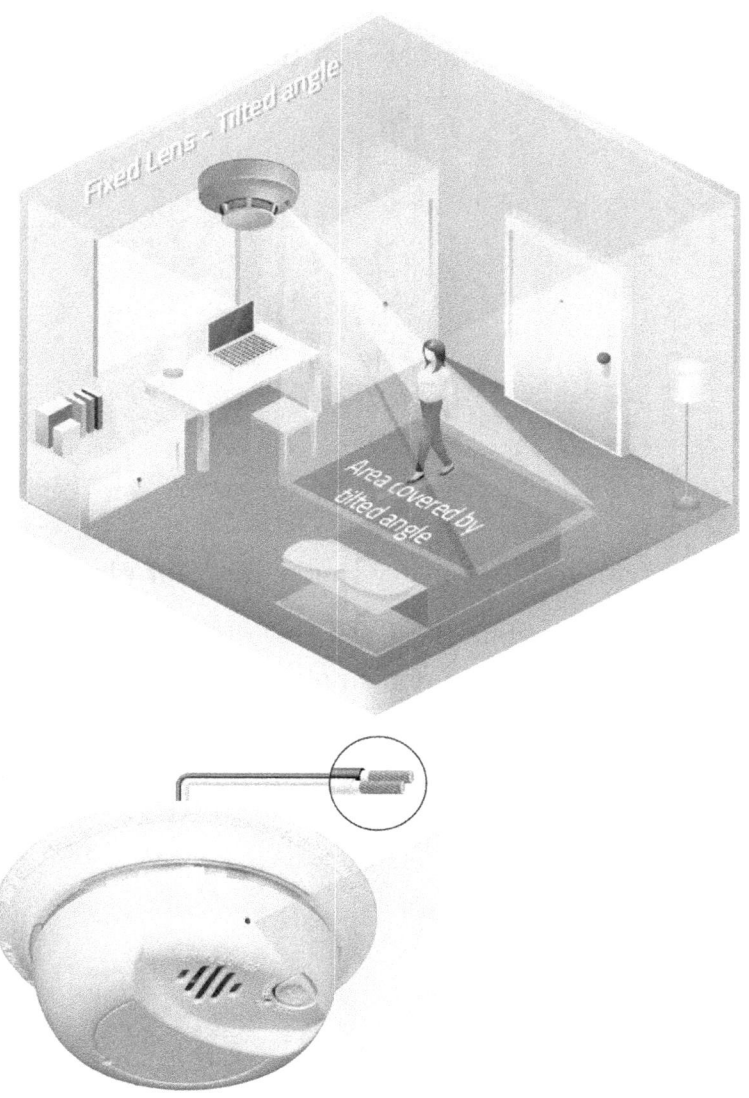

This type has the lens on the side of the smoke detector to cover a more significant angle of view instead of looking straight under the smoke detector. It can look at a tilted angle. For a side by side comparison between this type and the first type, please see below:

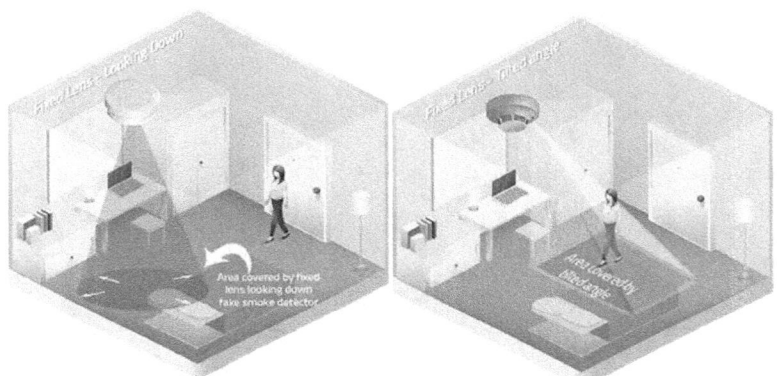

Because it's difficult to check every smoke detector physically, I have made a step by step guide for this specific camera to find out for sure whether it's a real smoke detector or a hidden camera.

<u>Step#1</u> Step out of the room and wait out of the sight of the smoke detector for at least one full minute.

<u>Step#2</u> Step into the room where the smoke detector (or the camera) can see you. Make sure you are in the area covered by the camera

If you are not sure where the area covered by the smoke alarm hidden camera starts, step out of the room completely and step back inside again.

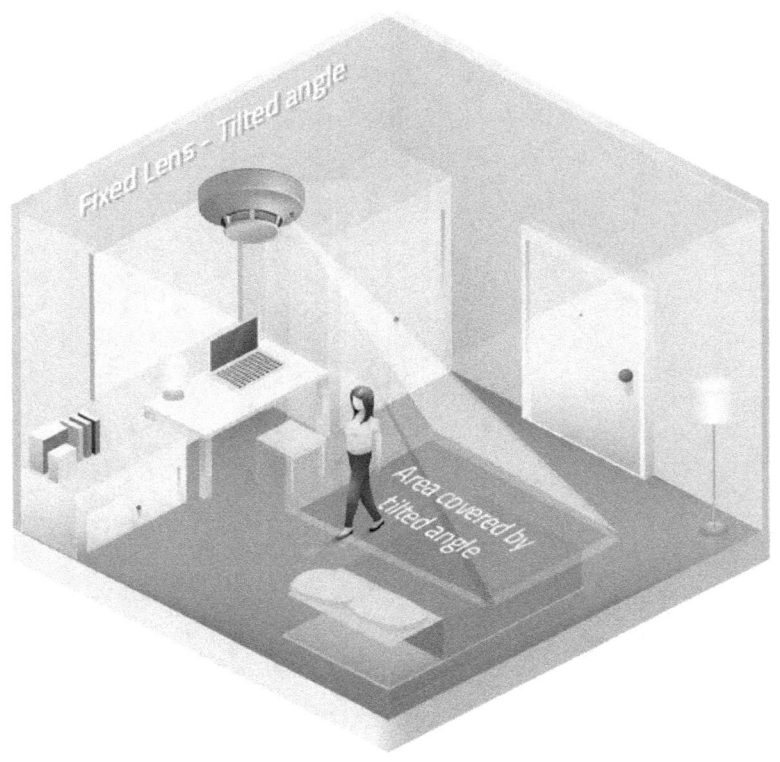

Step#3 Observe the smoke detector. If the smoke detector blinks in blue color for one time, this is a hidden camera.

Why?

Because most of the smoke detector hidden cameras available on the market use motion detection to save the space used for footage on the memory card. And they usually blink one time in blue once any motion is detected to indicate that recording has started. And this is how you know it's a camera, not a real smoke detector.

Type III: Rotating lens

This type of smoke detector hidden camera has a lens that can be tilted to cover a better angle of view as you can see below:

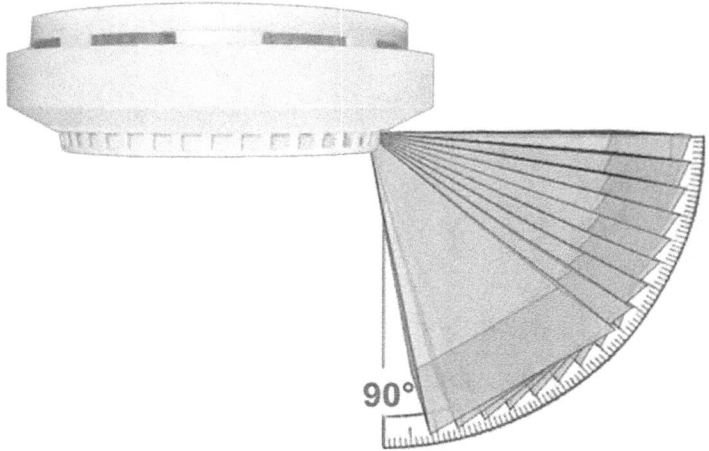

The owner of this hidden camera can rotate the lens before installing the smoke detector to the ceiling. This way he can adjust the angle of view specifically to the desired angle as shown below:

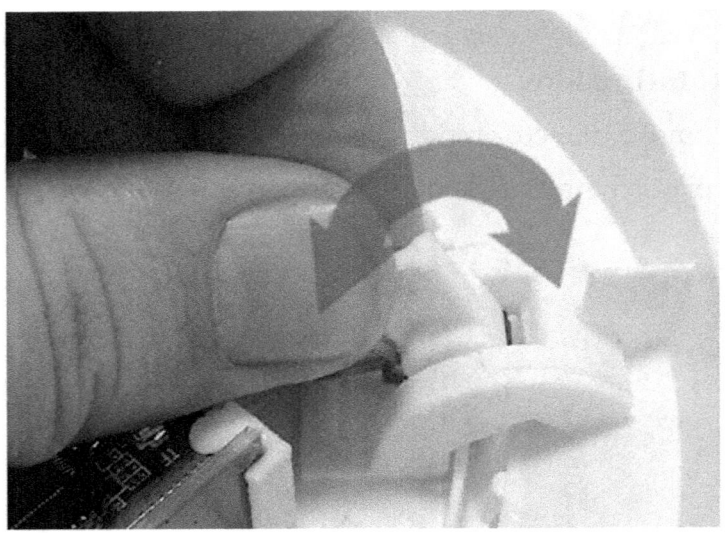

Again, all of these smoke detector hidden cameras share one thing, they all blink once (usually in blue) when any motion is detected. Therefore, they can be detected by the same procedure   mentioned   above.

Night Light Camera

Invisible Camera

Voice Control

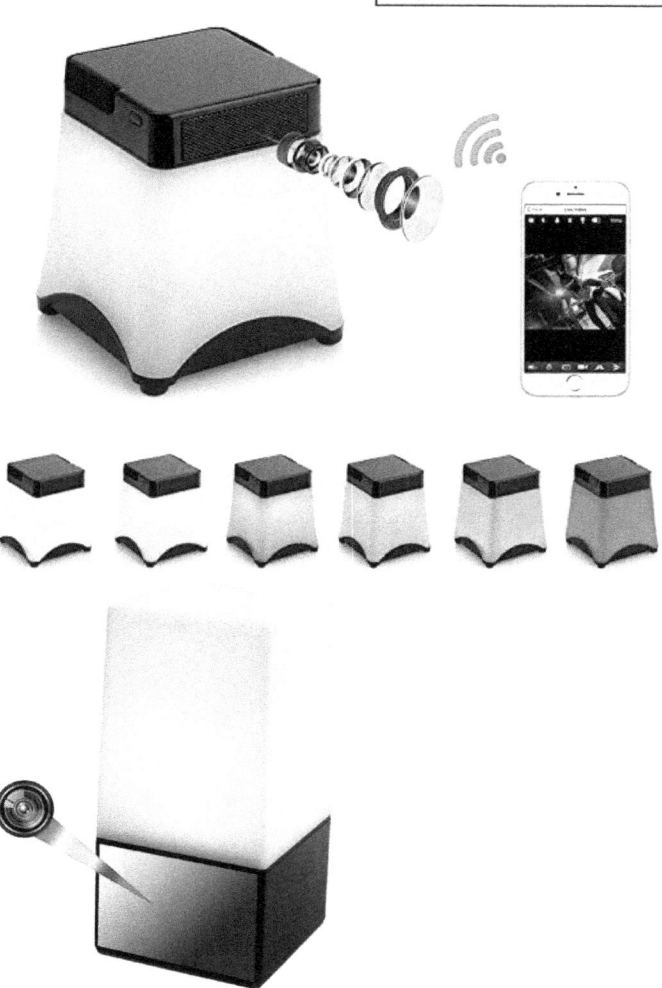

This night light hidden camera can come in two different shapes as shown above. The first shape has the camera above the light, and the second shape has the camera at the bottom of the nightlight. The nightlight itself can be lit in many different colors, and it needs to be connected directly to an AC outlet. The memory card is at the back of the unit.

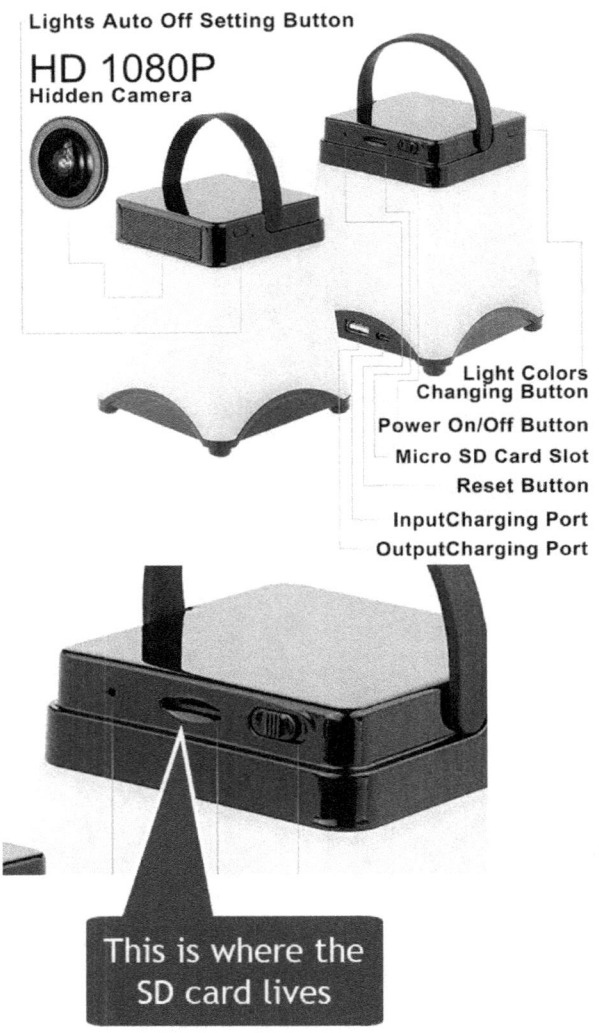

Lights Auto Off Setting Button

HD 1080P
Hidden Camera

Light Colors
Changing Button
Power On/Off Button
Micro SD Card Slot
Reset Button
InputCharging Port
OutputCharging Port

This is where the
SD card lives

This is considered one of the best-hidden camera footage quality wise. It can produce real full HD video footage, and it has a wide range of advanced functions such as very stable wifi connection and it support cloud uploading and local Micro SD card footage as well.

When you see one of the two shapes of night light shown above, flip the unit over and see the back of the unit to search for a memory card slot. As a fun, but useful side note, both variations of this camera are made by the same company as an OEM camera (meaning they sell it to other distributors and the distributors sell them under different brands and names) and once this camera is disconnected and reconnected to WiFi, it makes a loud sound saying "WiFi Connected!". So, you must have known by now that I will recommend disconnecting the wifi router if you are in a setup that allows doing so (your home or an Airbnb house, for example), then reconnect the router to AC and wit for "WiFi connected" coming from any suspicious device!

Power Strip Camera

Eight outlets, 4USB sockets and one hidden camera in one device.

8 Outlet +4 USB

Overload Protector
Protect household appliance at any moment

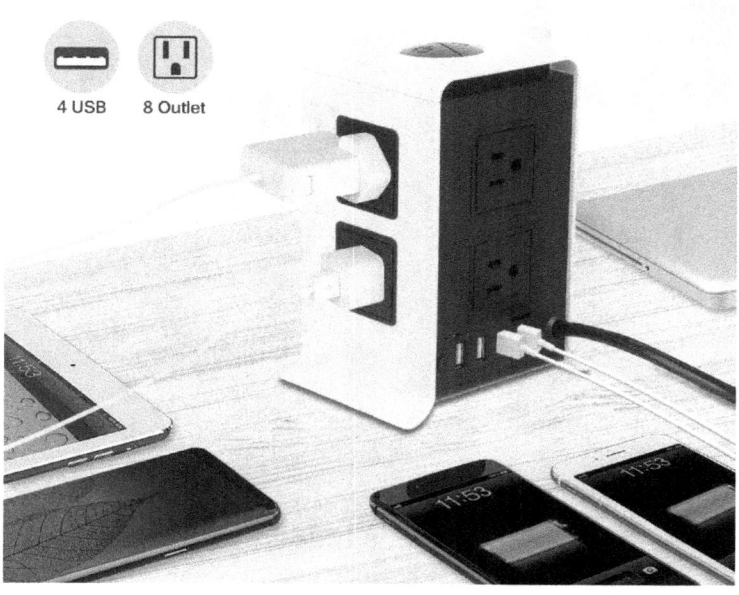

4 USB   8 Outlet

Electrical Outlet Camera

Camera

1080p 3.0MP FHD video

Things are getting creepy now! This hidden camera takes the shape of a normal AC electrical outlet.

The good news is, this camera is a non-functional AC outlet. Sooo, you can test it by connecting an electrical device to it and see if you are getting electricity! This can be done by connecting your phone charger to it and see if your phone is charging. The other fact about this model is it's installed using a sticky tape to the wall (the screws are fake!) so that's another way to spot this one.

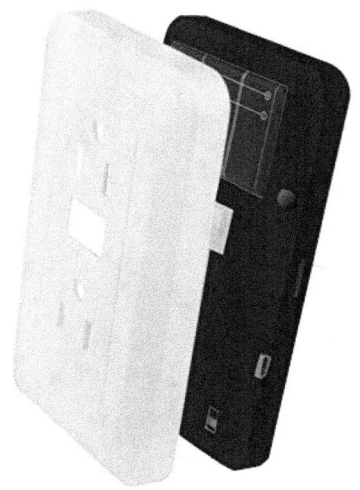

Keep in mind that the easiest way to spot this camera is to look for the sensor in the middle of the unit. No normal wall outlet has this sensor. The sensor is hiding behind a small semi-transparent glass or plastic and is necessary for determining the required exposure needed to shoot the video properly.

This camera is usually placed near the ground because the way it's designed is to look upwards at a 65-degree angle. Hence, the need to place the camera low.

The Light Bulb Camera

HD 1080P

360°

This light bulb has a black dot right at the center of the bulb that doesn't need to be there for a normal light bulb to function. I find it easy to spot because of that black lens. Although this one is self-explanatory, I wanted to include it in the book because placing this bulb in a very high position can make it difficult to notice.

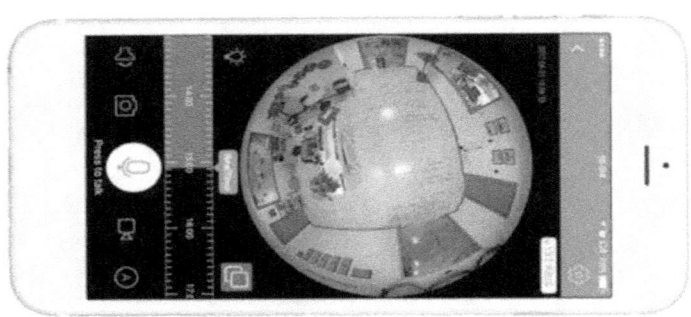

The Binder Camera

Well, I have to admit this is one of the more clever ones!

This binder has a hidden camera right at the center of the spine. It can hide perfectly in a library full of books

The micro SD card slot lives inside the book itself. Therefore, you will have to open the book to see the card slot as shown

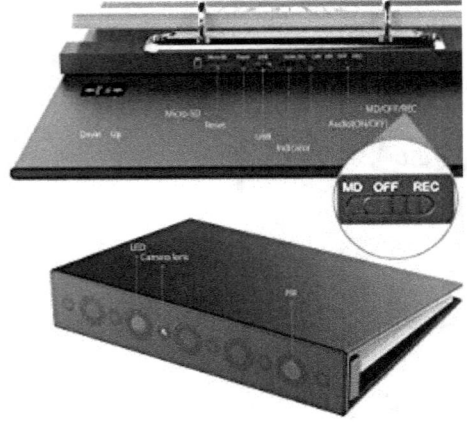

The Picture Frame Camera

This frame has a lens on the black frame, and it can be connected to the AC outlet directly as shown.

This wouldn't be very practical. Right? That's why this camera is capable of working without an electricity source for two hours. After the two hours, it will need to be recharged or plugged into an AC outlet.

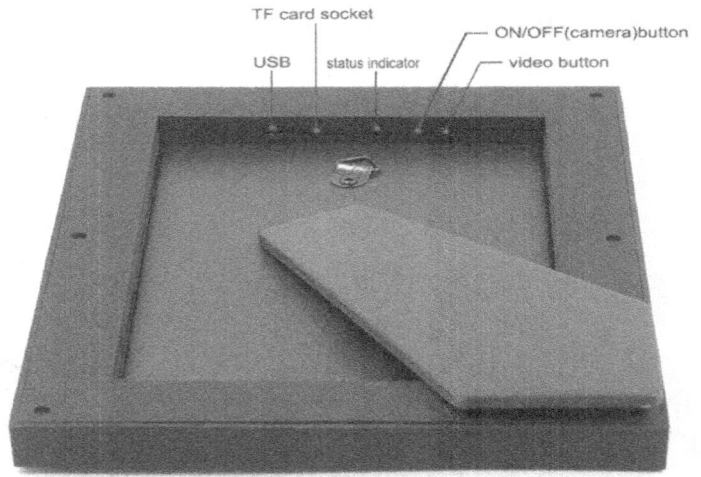

Phone Charger Dock Camera

It's a normal dock charging station for smartphones and perfectly does that function plus it doubles as a hidden camera that can loop record for 8hours then deletes the oldest video to record a new video. Dangerous is the right description for this device.

The back of the unit has a normal locking black piece of plastic that you have to remove to expose the memory card slot as shown below

And another variation of the same device:

This one is capable of wireless charging instead of direct contact cable charging. Still has a hidden camera though!

Bluetooth Music Player Camera

It's a nice Bluetooth player, and the sound quality is not bad! But that's not why people are buying this device. It's because of the 720p vivid spy camera hidden in this speaker.

And like most of the other hidden cameras, the memory card slot lives on the back side of the unit:

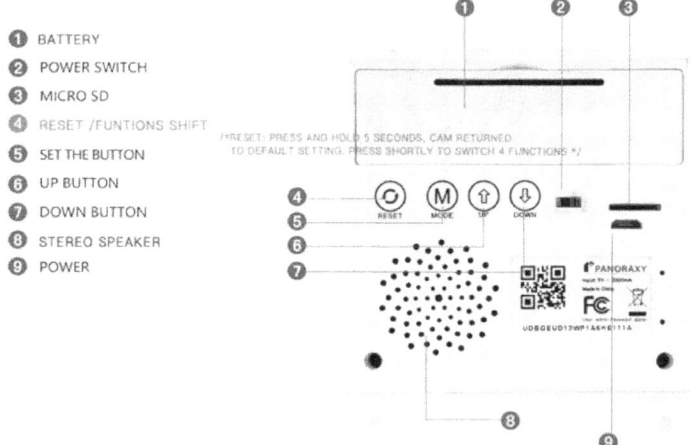

1. BATTERY
2. POWER SWITCH
3. MICRO SD
4. RESET /FUNTIONS SHIFT
5. SET THE BUTTON
6. UP BUTTON
7. DOWN BUTTON
8. STEREO SPEAKER
9. POWER

*/RESET: PRESS AND HOLD 5 SECONDS, CAM RETURNED TO DEFAULT SETTING. PRESS SHORTLY TO SWITCH 4 FUNCTIONS */

And another variation of the same idea:

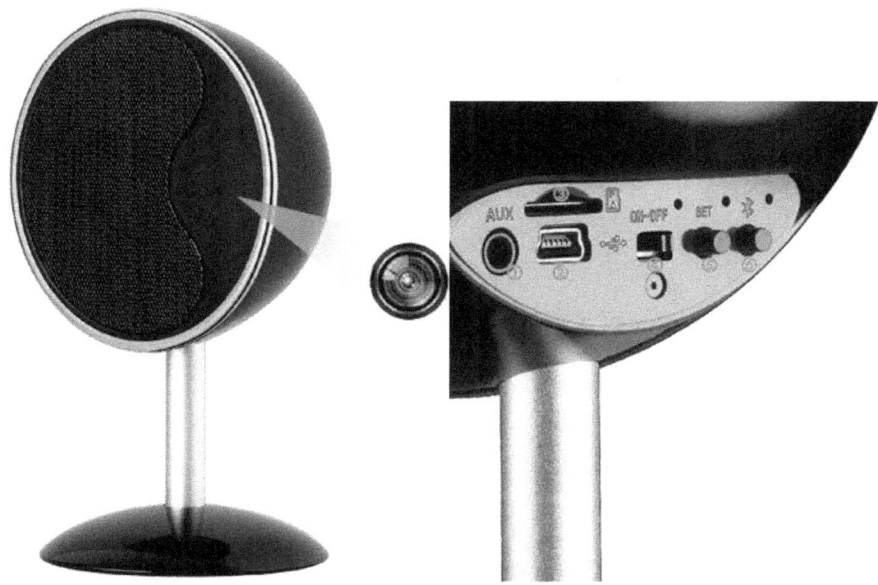

And more variations:

Water Bottle Camera

Yes, it does exist. It's not a legend. Behold, the water bottle hidden camera:

As you can see, the bottle has a void (groove) that fits a camera. And a sticker can be placed on top of the camera with a very small hole in the sticker just enough for the lens to see. After the bottle is used, the camera owner removes the sticker, offload the video footage, and replace the sticker with a new one.

6.3in

2.6in

0.68in

TF card | USB

LED

Camera

ON/OFF

1.49in

1.96in

Water

They may save the same sticker by sliding the sticker up or down, remove the camera from the special groove, and copy the footage.

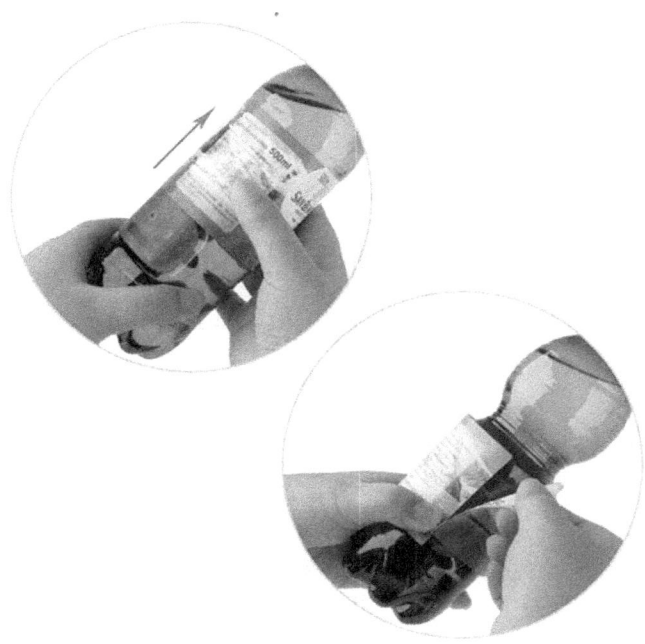

Because it comes with so many different fake labels (stickers) to hide the camera, you can technically use any brand of water or beverage to hide the camera as you can see below:

The bottle camera is one of the most unrecognized spy cameras on the market because it's an object no one ever suspects. So, the next time you see a bottle of water, you will know what to look for.

The Book Light Camera

This one is a bit smart, as I have used some of these USB book lights before. They are easy to use, and they can come in handy if you are a bookworm like myself! However, this one is not very innocent! See the pictures to see where the lens is located:

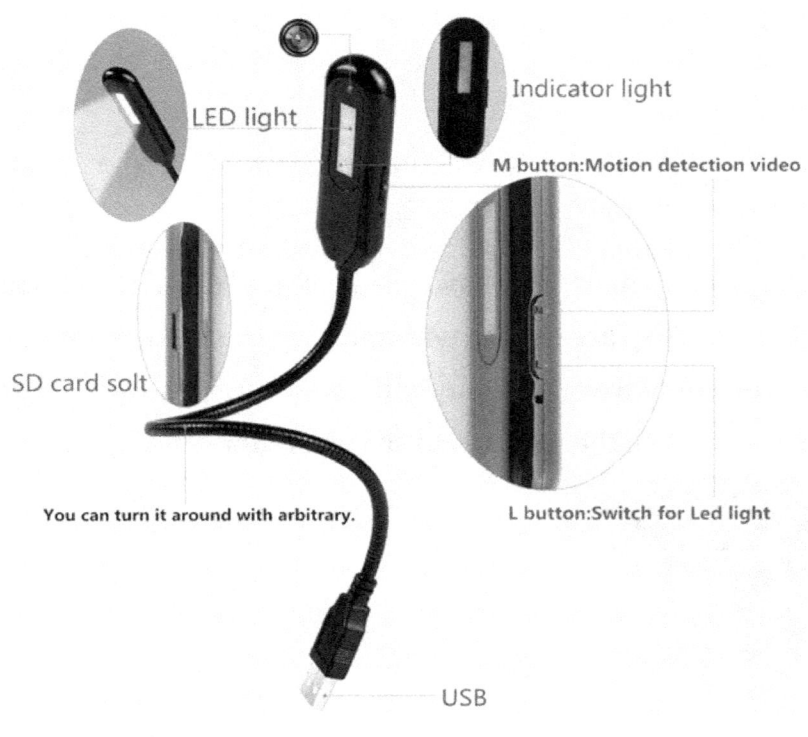

LED light

Indicator light

M button:Motion detection video

SD card solt

You can turn it around with arbitrary.

L button:Switch for Led light

USB

Chapter Five
"Wearable Hidden Cameras"

In this chapter, you will learn how the wearable hidden cameras look like, what shapes they may take and how to neutralize them!

The Watch Camera

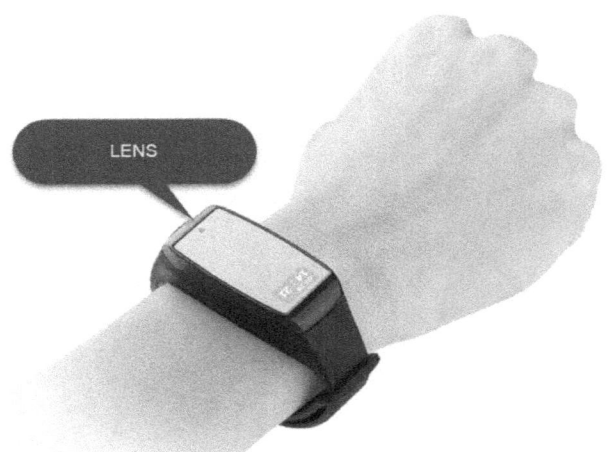

LENS

This smartwatch can sync with your smartphone and display notifications such as new messages plus it has a hidden camera that can record in the 1080p super clear resolution.

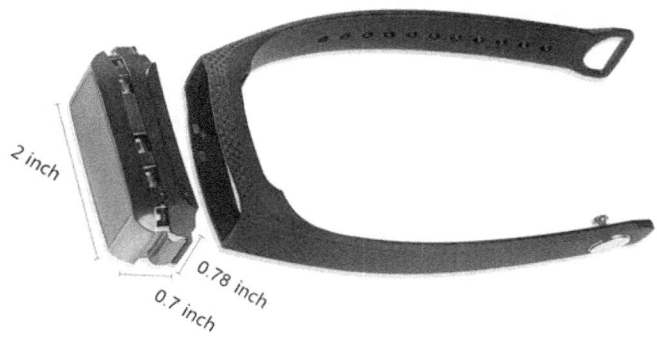

The memory card (Micro SD) is located on the side of the watch. But, you will have to take the watch body from the strap.

The Belt Camera

It's a fully functional belt and doubles as a hidden camera. See for yourself:

The memory card on this belt lives on the side of the buckle as shown below

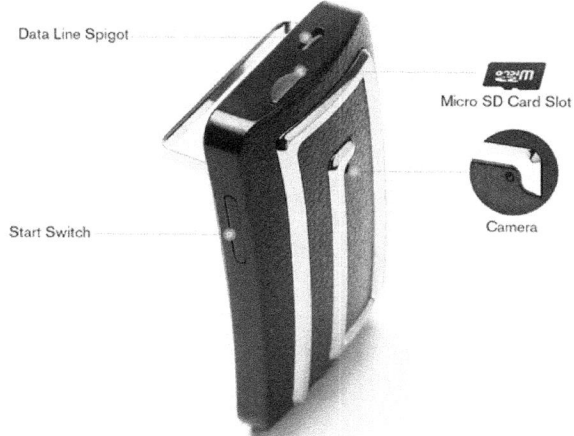

Data Line Spigot

Start Switch

Micro SD Card Slot

Camera

Keyfob Camera

Night Vision Indicator

Camera

The memory card lives on the side of the key fob and to expose it you need to press the key release button on the front of the key fob.

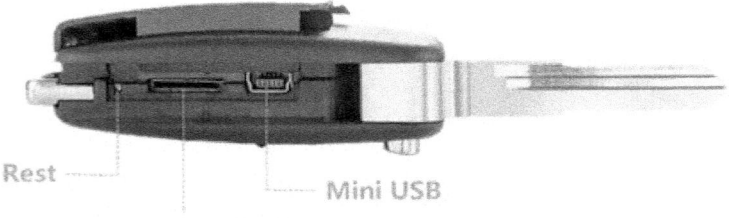

Rest

Mini USB

Micro SD Card Slot

Then Pen Camera

The lens opening can be covered by moving the pen cover which will also prevent the camera from recording.

It's a fully functional pen, and the lens is very well covered. And it starts recording with a single press on the top of the pen. It can record up to two hours.

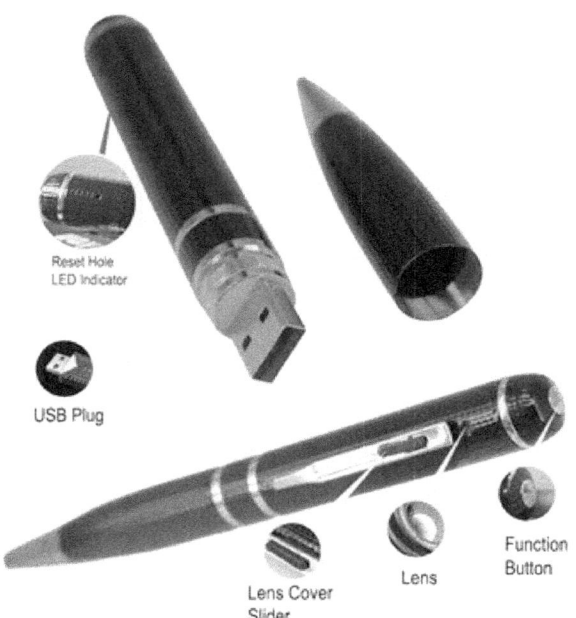

Reset Hole
LED Indicator

USB Plug

Lens Cover
Slider

Lens

Function
Button

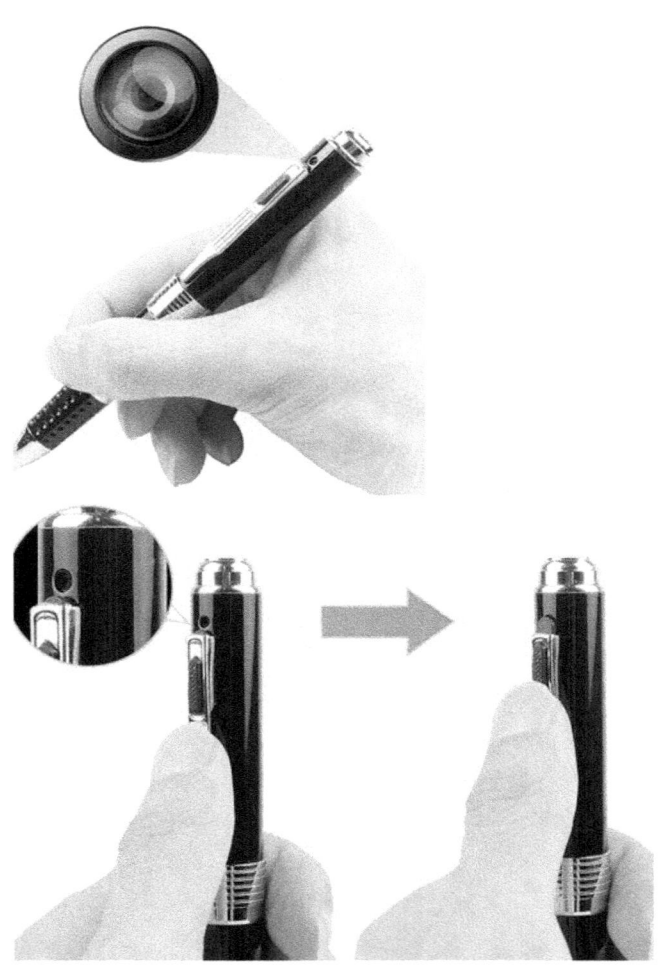

The micro SD card lives in one of the arms. I found it easy to spot this camera as the lens was not well covered. However, under direct sunlight, it might be easy to miss the lens.

The following is a better execution of the same idea (the lens is at the center of the glasses) however, it's more covert because the lens opening is smaller than the first one:

Keep in mind that this camera can be implemented into any glasses frame you can imagine!

The next variation of the same idea is much more difficult to spot because the lens takes the shape of one of the screws holding the glasses arms. And, another screw exists on the other side of the glasses which makes it difficult to spot:

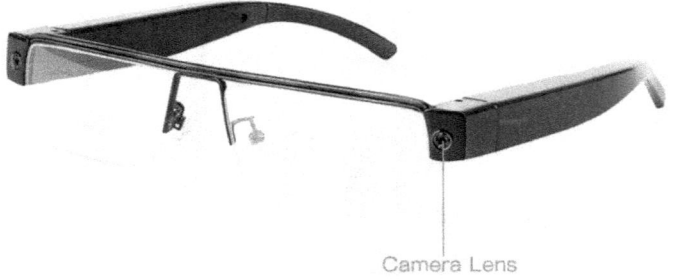

Camera Lens

So, in general: the camera on these glasses are either at the center front of the frame or on one of the arms to the right or the left.

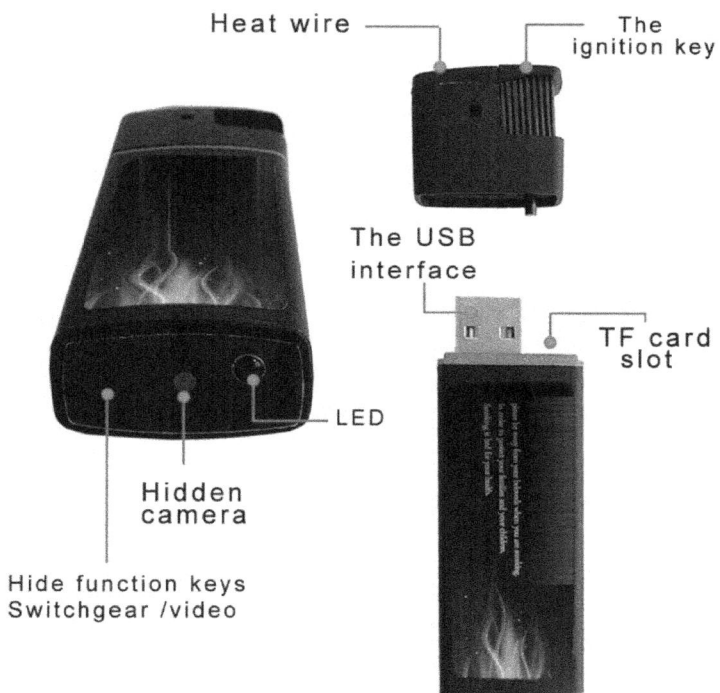

Lighter Camera

Heat wire —

The ignition key

The USB interface

TF card slot

LED

Hidden camera

Hide function keys
Switchgear /video

Not very practical because the camera is on the bottom of the lighter. So, if you notice that someone places his lighter on a counter with the bottom of the lighter facing you, you might be looking at a hidden camera.

Bluetooth Headset Camera

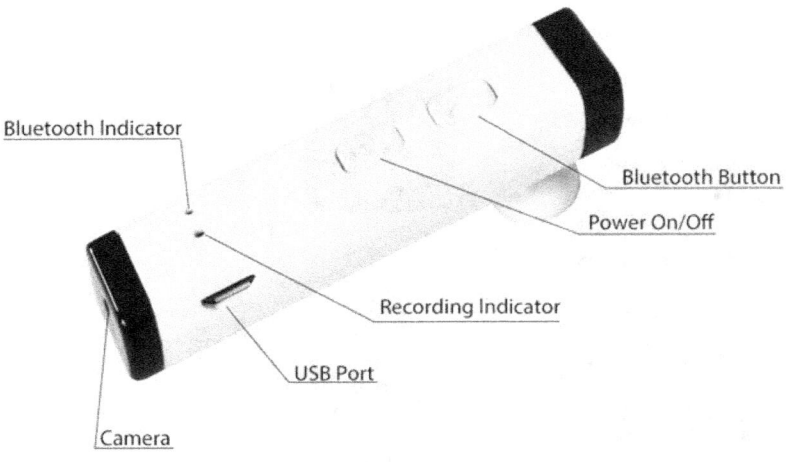

Bluetooth Indicator

Bluetooth Button

Power On/Off

Recording Indicator

USB Port

Camera

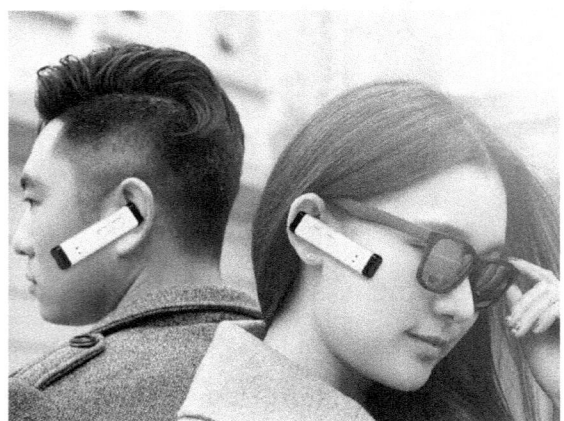

And more variations:

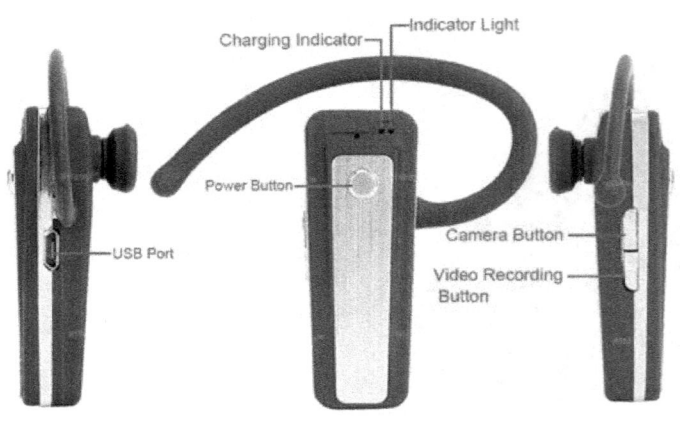

Charging Indicator — — Indicator Light

Power Button —

USB Port —

Camera Button —

Video Recording
Button —

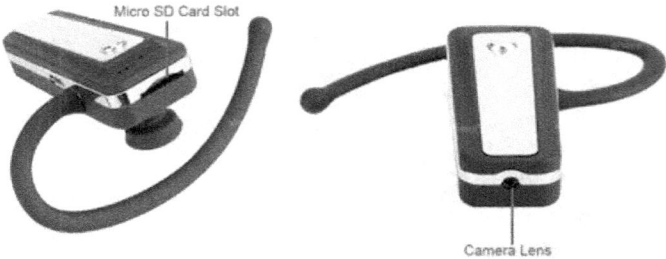

Micro SD Card Slot

Camera Lens

Hat Camera

First variation (and the easiest to spot):

Simple and very easy to spot even if you are not looking for it!

Second variation (and the hardest to spot)

The lens is very well covered by a small mesh of the same fabric used in the hat itself. So, it's very difficult to notice the lens unless if you know what you are looking for.

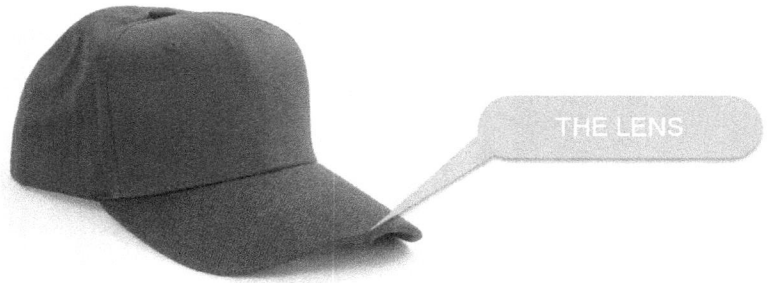

Keep in mind that hat cameras come in all shapes and colors! But they all share the location of the lens: either on the front tip of the hat (example#2) or the dead center of the head part (example#1).

Chapter Six
"Modular Cameras"

Analog Modular Camera

The reason modular cameras needed a separate chapter is that they are considered a more advanced level of spy cameras. If object cameras (indoors and wearable) are easy to use, they never provide the same quality as these modular cameras. Modular cameras are for someone very serious about capturing other people footage. This is the type of devices used by law enforcement authorities to bug someone's house. They need special skills to install, and they are extremely difficult to spot (as you can see below the freakin thing takes the shape of a screw!) And in addition to all of that, they record super vivid, super sharp footage and very reliable because they use wires and direct electricity source and they hide behind the wall!

15mm

15mm

800TVL

Analog

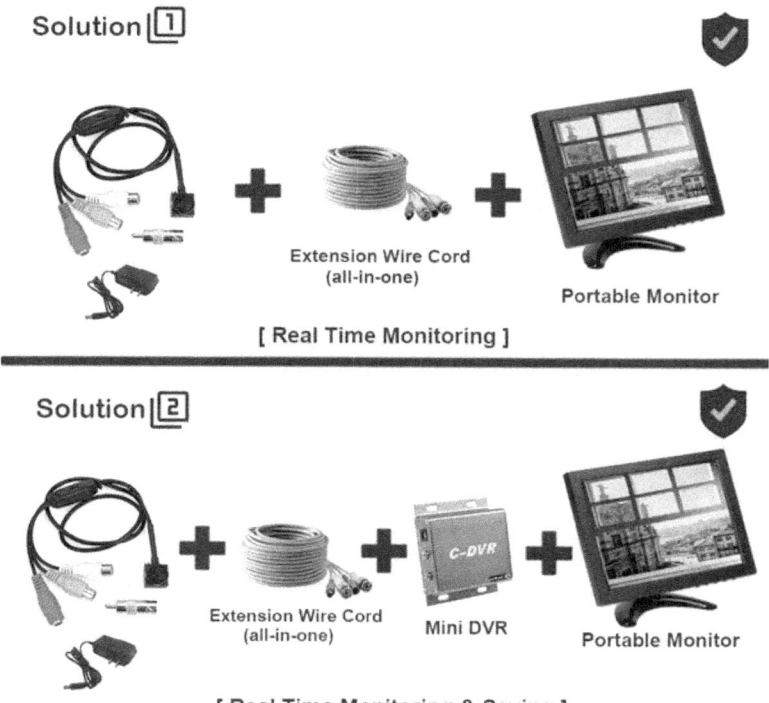

Solution ①

Extension Wire Cord
(all-in-one)

Portable Monitor

[Real Time Monitoring]

Solution ②

Extension Wire Cord
(all-in-one)

Mini DVR

Portable Monitor

C-DVR

[Real Time Monitoring & Saving]

The only way to detect this type of hidden camera is using a Radio Frequency Detector (RF Detector) as explained in Chapter Eight.

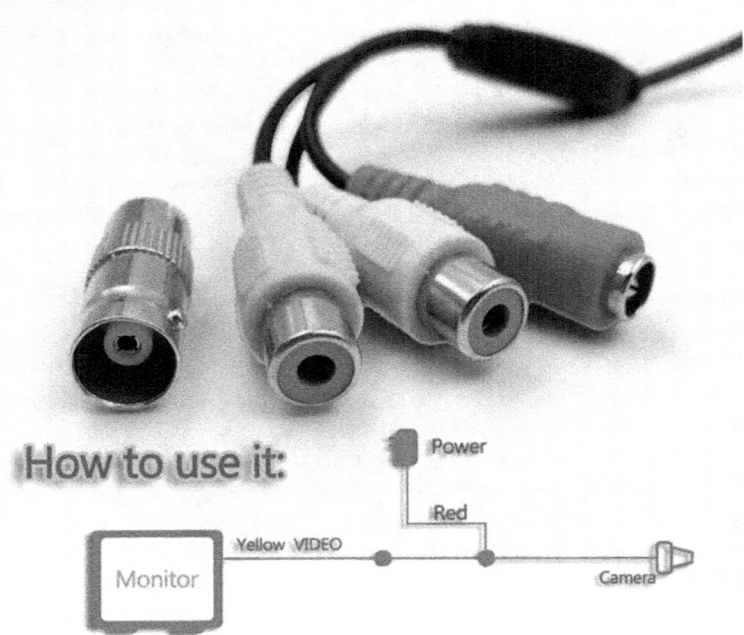

How to use it:

Monitor — Yellow VIDEO — Red — Power — Camera

Digital Modular Camera

HD

1080P

Support wifi

Motion Detect

Multi-User

Video

Camera

APP

Android / ios

Durable Folding
Any Angle

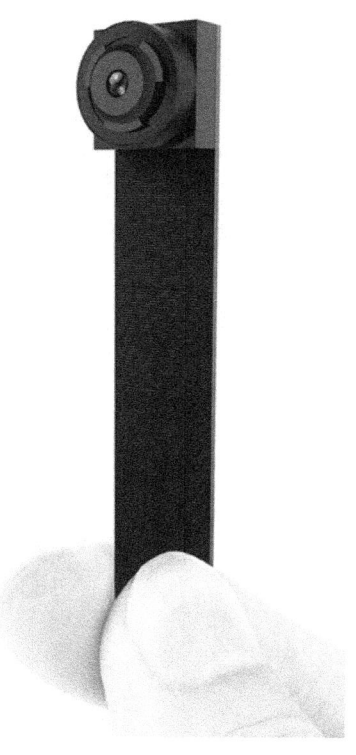

WIFI
Signal
Antenna

5.5in

package parts

Thin Battery

Portable Charger

USB AC Power Adapter

Micro USB 2.0

2.2in

0.9in

0.4in

Reset

OFF/ON

Micro
SD Card Slot

Reset   OFF / ON   Micro USB Interface

With USB Interface
Can be plugged in or connected to a power bank

Mobile charging   Power charging

Everything mentioned before about analog modular spy cameras applies on these digital modular spy cameras. They hide behind the wall, and they can use battery or direct wires to the electricity source, and they are super tiny, and they need an RF Detector to spot.

Chapter Seven
"Audio Recorders"

Every hidden camera mentioned in this book is capable of recording audio. And, because of that, there is no need to list all the cameras under this chapter as well. In this chapter, we will focus on the hidden audio recorders only. That means the devices listed here are not capable of recording video.

Super magnet

Sound Remote Monitor

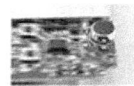 

Sensitive microphone + Audio DSP processor

Recording Time Standby Time

onetime charging   onetime charging

Automatic Recording Voice-
Activated Scheduled Recording

Easy Management Auto-Set file
name with setting file of recoder

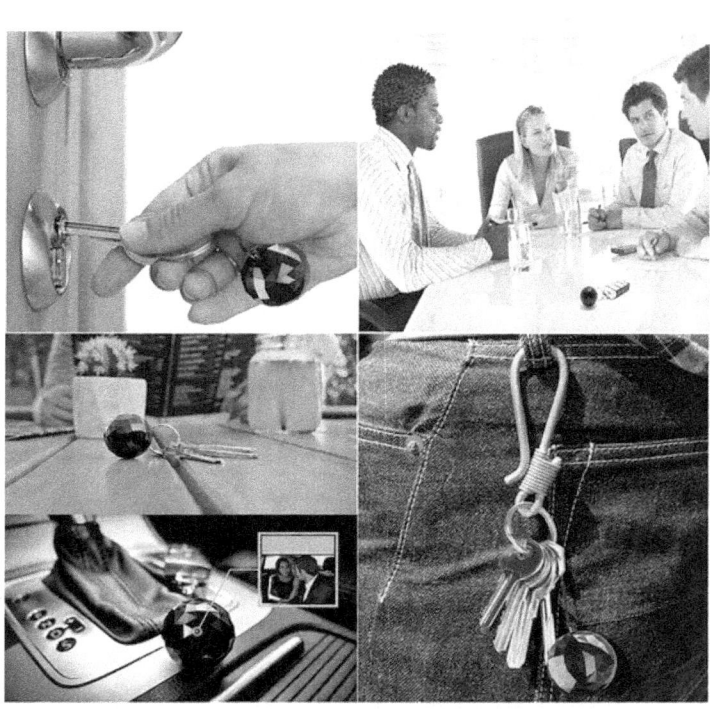

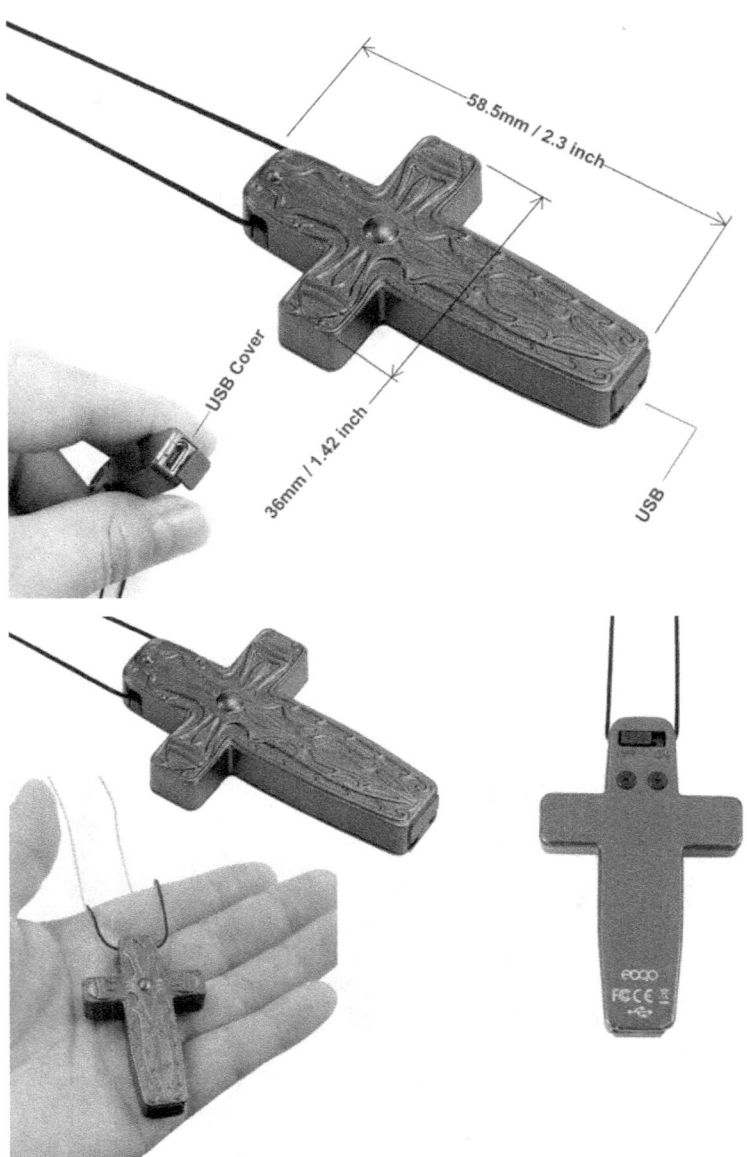

58.5mm / 2.3 inch

36mm / 1.42 inch

USB Cover

USB

Works as MP3 Player

Nacklace Voice Recorder

MP3 Player + Flash Driver

8GB Memory

One Button Start Recording

Chapter Eight
"Advanced Detection Techniques"

What if a new spy camera was released a few days after this book was published? The answer to this question is called RF Detector:

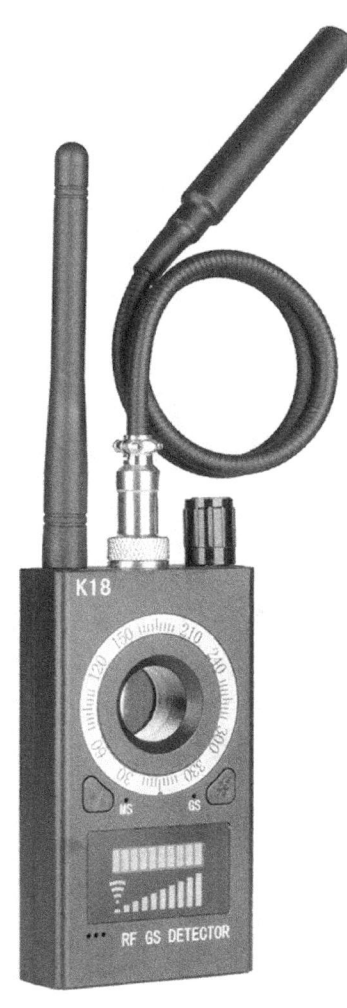

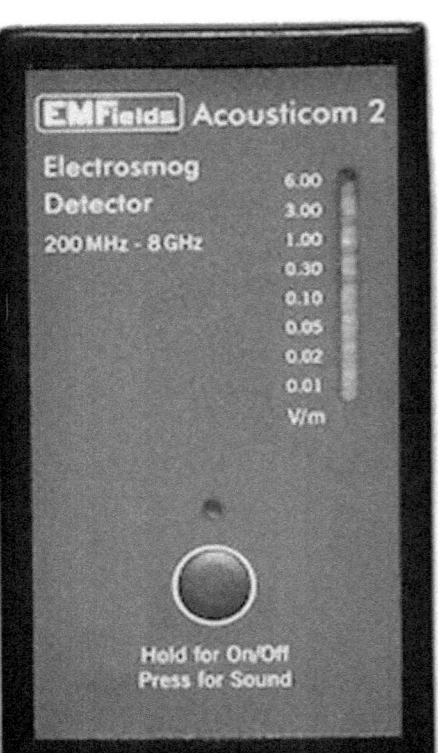

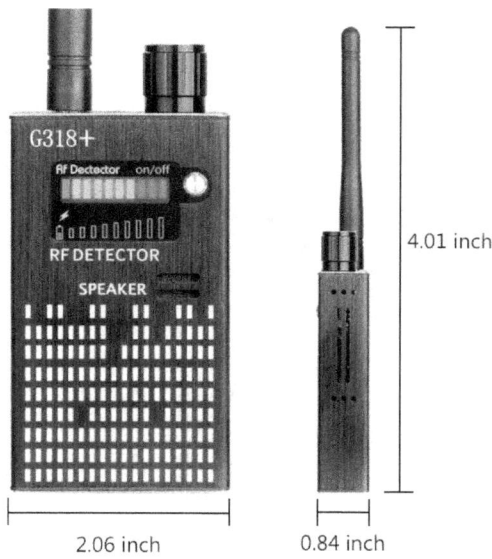

Frequency range: 1MHz-8000MHz
Detection dynamic range: >73Db
Indicating mode: 9 - level LED luminescence indicator variable tone voice indication
Power supply: built-in 3.7V1500mAH lithium polymer battery
Working current: 60mA
Continuous work: 15-25 hours

Detection sensitivity: 0.03mw (main frequency band)
Detection range:
2.4ghz wireless camera :10m^2(standard 10mW camera)
1.2GHz wireless camera :15m^2(standard 10mW camera)
Mobile phone signal 2G/3G/4G: 3-15m
Material: aluminum alloy
Weight: 160 g
Volume:4.01x2.06x0.84 inch

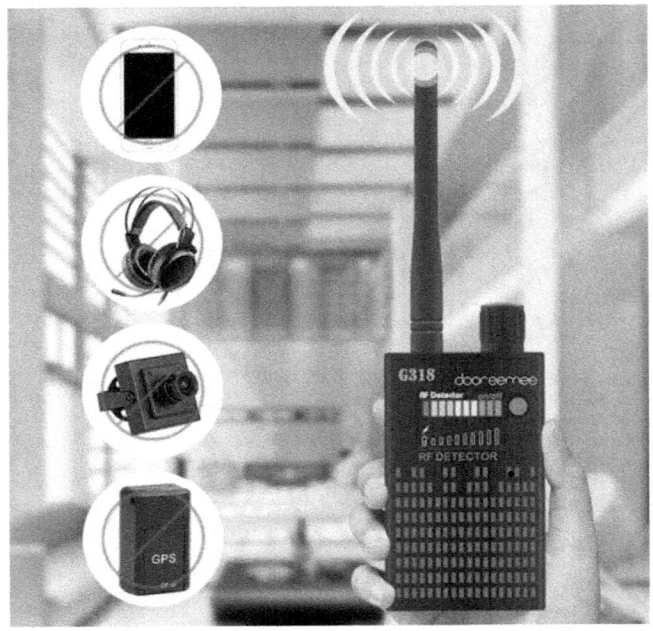

It's a small electronic device that detects radio frequency to detect a hidden camera when you put the detector near the camera. You turn the detector on and move around the area where you suspect a camera may be hidden. Once the device detects a signal, it will make a sound, and the lights on the front will turn on. The RF Detector comes in many shapes packed with many different features. The more sensitive the detector, the pricier it gets.

Almost every RF detector share two main functions: 1. Radio signal frequency detector. As explained above the device will make a sound when radio signals are detected. 2. Red light detector. As you can see from the pictures above, the device has a hole in the middle covered with red plastic. The detector will send strong red light signals, and when you look through that hole, the red light reflects from the lens of the hidden camera which makes it clear to see the lens of the hidden camera through that red hole!

As mentioned before, pricier RF detectors may have more functions such as hidden audio recording device detection (bugs). And GPS device detection and even spy SIM-card detection. But, they cost more than the average basic RF Detector.

Because it's difficult to demonstrate how to operate such device by pictures only, I have included a bonus video for free for the book readers that shows how to use the RF Detector step by step; this video can be found under the bonus section on the book's official website https://suspicious.tech/bonus/

To get yourself an RF Detector for a discounted best price exclusively for the book readers, visit https://suspicious.tech

Chapter Nine
"Now What?"

The question you will find yourself asking after spotting a hidden camera is: Now What?

Let's be honest, hearing about someone who was spying on using a hidden camera is one story. And being in the actual situation as the victim of video voyeurism is a whole different story.

One of the victims of video voyeurism described the moment they discovered a hidden camera recording them by saying "I felt paralyzed!. I didn't know what to do, although I always had a feeling something wrong with that place."

This was a very accurate way to describe the feeling of your privacy being invaded by video voyeurism. Even if you know everything about hidden cameras, the first time to experience being spied on is always a shocking moment.

Here's what you need to do:

1. Call the police.

   Even if you can destroy the camera or the spying device and assuming that you were able to destroy all the footage. It's still highly recommended to call the police. This way you have some sort of guarantee that the person who is using spy cameras to illegally spy on people doesn't blackmail you later.

2. Don't leave the crime scene.

   Yes, it's a crime scene if you haven't realized that by now. If you leave the crime scene the camera owner may hide it or destroy the evidence until the police arrive. If it's a hidden camera in a fitting room of a clothes store, there is a very high chance that the store personnel will try to cover up and will try to calm you down and to escort you to a different room other than the bugged room to get rid of the evidence until the police arrives.

Terminology

AC Outlet: Electricity source

SD Card: used in the book to refer to either Micro SD card or full-size SD card. A memory card to save all the video files recorded by the camera.

Electrical Outlet: Same as AC Outlet.

The Cloud: When a camera is capable of saving all the recorded footage to the cloud, that means the camera will simultaneously upload the footage to an online account, and the footage can be retrieved from that online account only using the owner's predefined username and password.

Resources

An easy way to find the latest hidden/spy cameras released is to visit various online stores and search for the term "Hidden Camera."

The following link will give you easy access to the latest hidden cameras we found listed according to the time released:

https://suspicious.tech/bonus

Also, www.tgwa.tv is a great source of up-to-date news related to hidden cameras and spy devices.

Conclusion

Thank you again for reading this book!

I hope this book helps you to spot hidden camera and protect your privacy. The final advice I can give you before you close this book is to have a skeptical eye and never trust any device even if that device is not mentioned in this book. The rule is, if it has power (electricity), it may be a hidden camera. Thank you and see you in the next book!

Bonus Section

As a valued member of the book family. You get a unique code to unlock a bonus section on the book's official website. That bonus section contains exclusive video tutorials, tips and tricks all related to enhancing your security and privacy.

The Bonus Section can be found on www.suspicious.tech/bonus

Your Bonus Section code is 24FRDGTWES

Also under the bonus section, you can request your PDF copy of the book signed by me and sent to your email. All you need to do is to visit the bonus section and use the form to send your name, and I will send you a signed PDF copy of the book with colors (Kindle devices show any book in black and white only, and Kindle for Android and iOS can show colors).

Please don't share your bonus section code with others as this may revoke your Bonus Section access privileges.